Successful Learning Simplified

A Visual Guide

Fiona McPherson, PhD

Published 2016 by Wayz Press, Wellington, New Zealand.

Copyright © 2016 by Fiona McPherson.

All rights reserved.
No part of this publication may be reproduced, stored in a retrieval system, or transmitted in any form or by any means, electronic, mechanical, recording or otherwise, without the prior written permission of Wayz Press, a subsidiary of Capital Research Limited.

ISBN 978-1-927166-20-8

To report errors, please email errata@wayz.co.nz
For additional resources and up-to-date information about any errors, go to the Mempowered website at www.mempowered.com

Also by Dr McPherson

Easy Russian Alphabet: A Visual Workbook

Indo-European Cognate Dictionary

Mnemonics for Study (2nd ed.)

My Memory Journal

How to Approach Learning: What teachers and students should know about succeeding in school

How to Learn: The 10 principles of effective practice and revision

Effective Notetaking (3rd ed.)

Planning to Remember: How to remember what you're doing and what you plan to do

Perfect Memory Training

The Memory Key

Contents

About this book	1
Introduction	2

How memory works — 5
The nature of long-term memory	6
The importance of working memory	8
A lot of editing goes on!	11
Review questions	13

Approaching your learning — 15
Priming	16
Goal setting	17
Evaluating the text	18
How blood flows	20
Early America	21
The role of consolidation in memory	26
Ozone & UV Radiation	29
Introducing brain cells	32
Human gene affects memory	34
Review questions	38

Getting an understanding of the text — 41
Summaries	42
Types of summary	42
Topical summaries and overviews	43
Advance organizers	44
Skimming	45
Headings	46
Identifying text structure	51
Illustrations	55
Graphics	62
Checklist	68
Text	70
Main points	74
Review questions	76

Reading — 79
Levels of processing	80
Active reading	83
Mental models	83
Reading expository texts	85
How narrative & expository texts differ	86
Approaching expository text as a novice	88
Active reading strategies	89
Main points	90
Review questions	91

Taking notes — 93

- Notetaking makes information meaningful — 94
- Highlighting — 95
- Creating summaries — 100
  - How to deal with different types of text structure — 104
  - Summarizing isn't just about the notes you take — 112
  - Outlines and Graphic Organizers — 112
  - More graphic organizers — 114
- Main points — 120
- Review questions — 122

Learning through understanding — 125

- Understanding is rooted in the connections you make — 126
  - Asking questions — 126
  - Making comparisons — 129
  - Analogies — 129
- Concept maps — 131
  - Mind maps — 135
- What this means for notetaking — 137
- Review questions — 139

Reading on the web — 141

- Navigating a hypermedia environment — 142
- Problems & benefits of animations — 146
- How to learn effectively in hypermedia environments — 147
- Review questions — 149

Getting the most out of lectures — 151

- How lecture notes are different from textbook notes — 152
- Are there special strategies for taking notes in lectures? — 153
- Different approaches to lecture notetaking — 155
- Main points — 156
- Review questions — 157

Memorizing verbatim — 159

- Mnemonics — 160
  - Vocabulary — 160
  - Order — 161
  - Words vs images — 162
- Rhythm & rhyme — 162
- Keyword method — 163
- First-letter mnemonics — 166
- Simple list mnemonics — 168
  - Story or sentence mnemonic — 168
  - Link mnemonic — 169
- More complex list mnemonics — 170
  - Method of loci — 170
  - Pegword mnemonic — 171
  - Coding mnemonics — 172
- Main points — 175
- Review questions — 176

Revising	**179**
Types of retrieval practice	182
Flashcards	182
Q & A	183
Re-summarizing	186
How often do you need to practice?	188
Spacing your practice	188
Spacing within your study session	188
Spacing between review sessions	190
The ten principles of effective practice	192
Review questions	193

Skills	**195**
Critical factors in mastering skills	196
The ten principles of effective skill practice	198
Review questions	199

Subjects	**201**
Reading	202
Mathematics	202
Science	203
History	203
Note-taking	203
Writing	204
Science report	205
History report	206
Philosophy essay	207
Understanding	208
Remembering	208
Foreign Languages	208
English literature	209
History	209
Mathematics	209
Science	209
Main points	210
Review questions	210

Putting it all together	**211**
The main steps	212
Your toolbox	213

Glossary	**214**

Answers to Review Questions	**220**

About this book

Mempowered Visual Guides are designed to present the essential information in as clear and readable fashion as possible. The focus is on these three words: essential; clear; readable.

This Visual Guide is supported by three study skills books that provide the detail and research backing that isn't covered here:

> *How to Learn*, for information on how to revise and practice
>
> *Effective Notetaking*, for information on learning from textbooks and lectures
>
> *Mnemonics for Study*, for information about memorizing facts.

For more information about how memory works and how these principles relate to specific memory strategies, go to *Perfect Memory Training* or *The Memory Key*.

This Visual Guide can be used as a quick reference to augment these books, or as a simple and direct guide to the essential principles, without the explanations and more in-depth discussion provided in those books.

Introduction

There are many paths to success in study. For a lucky few, success may come from natural talent. That's one path. But for most, it's all about being a *smart user of effective strategies*. The strategies you use, and the way that you use them — these constitute all the other paths to successful study.

This is the key point: it's not (only) about intelligence; it's not (only) about hours put in. Sure, it helps to be smart, and it helps (even more) to be diligent. But many students have wasted a great many hours on study that is simply badly directed. What is important is using effective strategies effectively.

To use a strategy effectively you need to understand something of how it works. When you understand this, you will know when to use any given strategy, and when not to use it. Because this is the critical point that few students understand: it's not enough to have one good strategy. No single strategy is the right one in every learning situation.

To study effectively, you need a tool-box.

That's what this book is about — giving you that tool-box.

Here's your first tool.

Whenever you approach a learning experience, be it a textbook or a lecture or some multimedia web-type resource, you get off to the best start by preparing your mind. In cognitive psychology, this is called **priming**. Like priming the pump, your mind works more effectively if you've activated the right parts of it first.

So, to enable you to get the most benefit from this book, let's start by getting you primed. Stop a moment and reflect on how you learn. Here are a few questions to help you think.

When you're given specific material to read (a textbook chapter; an article; a handout), how do you read it? Do you

- skim it quickly
- read it once, from beginning to end
- read it carefully, going back and forth
- read it repeatedly
- take detailed notes
- take linear notes
- make a **mind map**
- …?

When you go to a lecture, do you

- prepare for it by reading relevant material beforehand
- think about the topic beforehand
- think up questions you hope to have answered
- try to take down everything the lecturer says
- try to capture only the important information
- sketch out the main points in a loose diagram
- organize the information as you hear it
- use your own words when taking notes
- re-do your notes afterward, by printing them out neatly or typing them up
- re-work your notes afterward into a more organized and meaningful format
- re-phrase your verbatim notes into your own words
- …?

Do you remember best what you *hear* or what you *read*?

Do you find it easier to understand something if someone explains it to you or if you study a written text?

How do you try to memorize facts, such as lists or new words? By

- repeating them over and over
- testing yourself with flashcards
- writing them out
- using **mnemonic** strategies
- ...?

When you read a textbook, do you

- read only the main text
- give due attention to any diagrams and illustrations
- attend to any summaries first
- read summaries last
- ...?

I hope this has got you thinking about your study skills, and that you are now ready to learn about learning (and which of these strategies are worthwhile).

As I said, effective learning — learning that lasts; learning that is achieved efficiently, with little waste of time or effort — cannot be achieved with a single strategy, and to know when and how to use which strategies, you do need a very basic knowledge of how memory works. Let's begin, therefore, with a quick sketch of the essentials of memory.

How Memory Works

Understanding how memory works is the first step to improving your memory and learning skills.

When you understand how and why specific strategies work, you'll know when and how to use them, and how to modify them to fit the situation and to suit you. You'll also remember them better, and be more likely to use them.

The nature of long-term memory

Many people think of memory as a recording, but this is not simply an error — it is a dangerous error.

The essential nature of memory is that it is:

> coded

> heavily edited

> made of connected items that you re-assemble into a coherent picture every time you 'remember' something.

Long-term memory is a network, and every individual memory in it is a (much smaller) network within it.

Our memories are held in connected nodes of the network.

These nodes are activated by signals passing along the 'paths' that connect them.

These signals begin in response to a **retrieval cue** (a question, a thought, a perception, …) — anything that triggers retrieval of a memory (i.e., "remembering"), as in the simplified network below.

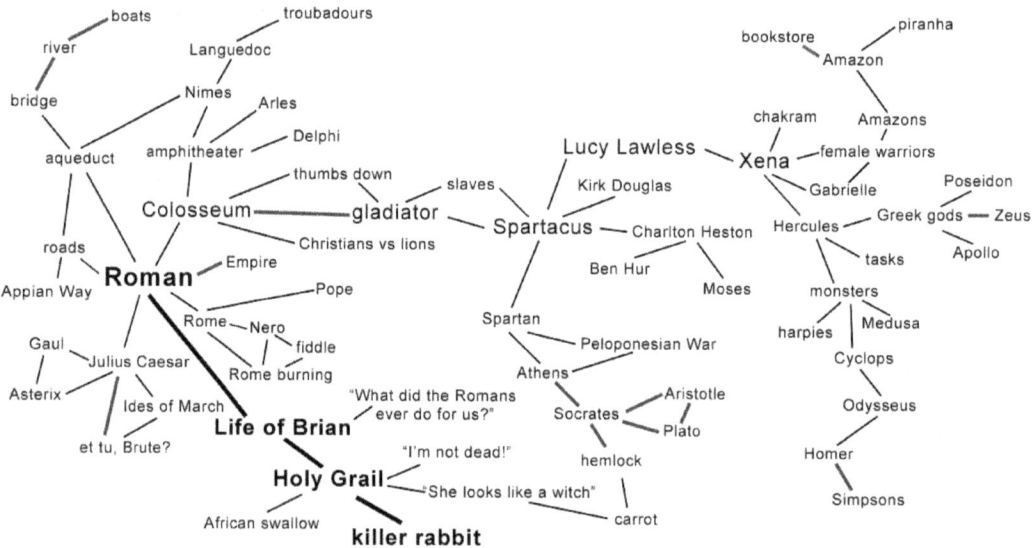

From these basic facts come 8 basic principles of memory that are vital for understanding how to improve your memory and learning:

1. **code principle**: memories are selected and edited codes (this is why putting information into memory is known as **encoding**).
2. **network principle**: memory consists of links between associated codes.
3. **domino principle**: the activation of one memory code triggers connected codes.
4. **recency effect**: a recently retrieved code will be more easily found.
5. **priming effect**: a code will be more easily found if linked codes have just been retrieved.
6. **frequency (repetition) effect**: the more often a code has been retrieved, the easier it becomes to find.
7. **matching effect**: a code will be more easily found the more closely the retrieval cue matches the code.

    This can be seen in jokes: if you were asked, "What did the tree do when the bank closed?", you'd probably realize instantly that the answer had something to do with "branch", because "branch" is likely to be a strong part of both your "tree" code and your "bank" code. On the other hand, if you were asked, "What tree is made of stone?", the answer (lime tree) is not nearly as easily retrieved, because "lime" is probably not a strong part of either your "tree" code or your "stone" code.

8. **context effect**: a code will be more easily found if the encoding and retrieval contexts match.

    If you learned about Henry VIII from watching The Tudors on TV, you'll find it easier to remember facts about Henry VIII when you're sitting watching TV. We use this principle whenever we try and remember an event by imagining ourselves in the place where the event happened.

The importance of working memory

Our memory is huge, but like an iceberg, most of it is hidden from view. At any moment you are only aware of the tiny piece of it that you're thinking about.

Working memory is that very very small part of memory that you're consciously aware of. It's the place where you turn new information into memories, where you put memories you've retrieved from long-term storage so that you can 'look' at them, where you put information together and think about it. Working memory is your mental workspace.

Working memory governs your ability to:

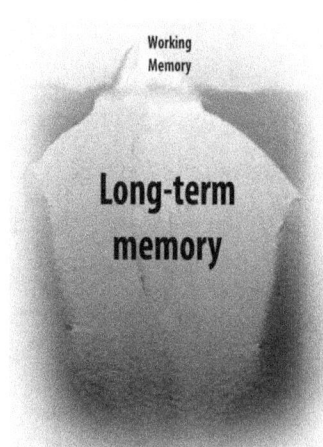

- understand what you're reading or hearing
- learn new words
- reason
- calculate
- plan and organize yourself

and much more.

Working memory capacity (WMC) — the amount of information you can hold in working memory at one time — varies a little between people. While the amount of variation isn't large, it does have a significant impact on your cognition.

The size of your working memory affects (among other tasks):

- your ability to take good notes
- your ability to understand complex text
- your ability to make inferences
- your ability to learn a new language.

Students with a low working memory capacity tend to make their problems worse by:

- using less effective strategies
- not practicing enough (because they see other people don't need so much practice and don't realize they need to compensate for their lower WMC)
- worrying (worrying uses up some of your working memory, so you're effectively reducing your WMC even more).

You can increase your WMC by:

- using both working memory stores (e.g., repeating some of the information you're trying to hold in mind while holding other information as a visual image)
- speaking or thinking quickly (you can only hold in verbal working memory what you can say in 1.5 — 2 seconds)
- increasing the size of your chunks
- reducing distraction.

> ### What working memory is
> Working memory is the mental space where we actively and consciously process information.
>
> Working memory is the gateway to your long-term memory.
>
> The size or capacity of working memory varies a little between people, but is always very small: 3 to 5 chunks.
>
> There is more than one working memory. One space is for processing verbal information; another for processing visuospatial information.
>
> There is also a controller, that controls where you direct your attention and how well you can resist distraction, switch between tasks etc.

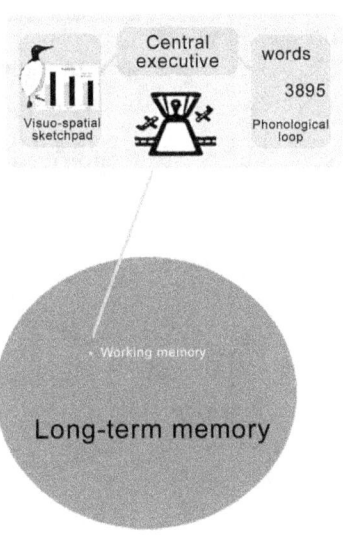

A **chunk** is information so tightly connected that it can be treated as a single unit. For example, these nine words:

1. brown
2. the
3. jumps
4. dog
5. over
6. quick
7. lazy

How Memory Works

8. the

9. fox

could be nine chunks, or, in a different order ("the quick brown fox jumps over the lazy dog"), one chunk (for those who know it well as an example used in typing practice). At a much higher level of expertise, a chess master may have whole complex sequences of chess moves as single chunks. The construction of complex chunks is one of the foundation stones of expertise.

The third working memory concept that is critical to your application of the appropriate study strategies is that of **cognitive load**.

Cognitive load is the extent to which a situation makes demands on your working memory.

Cognitive load depends on:

> - the difficulty of the information
> - how the information is presented
> - your existing knowledge
> - your WMC
> - the distractions in your environment and in your mind (other thoughts, concerns).

To a large extent, then, cognitive load is reflected in task difficulty (that is, how difficult you find the task).

Here are some factors to consider when assessing cognitive load:

> - **how much information there is**
> - **how complex the information is**
>
>   How difficult it is to craft it into a tightly-bound cluster — that is, how hard it is for you to connect the information together and make sense of it.
>
> - **how often you have to shift your focus**
>
>   For example, if you're trying to do more than one task at a time (writing an essay and checking your Facebook page), then you are appreciably adding to your cognitive load; if you are trying to translate a text in a foreign language and have to keep diverting to

the dictionary and a grammar book, then this will add considerably to the cognitive load (as compared to being able to translate without needing to check vocabulary and grammar points).

- **whether information already in focus has to be altered and how much time and effort is needed to change it**

    For example, if you are mentally calculating the answer to (38 x 4)/8, you will need to keep 'updating' the number in your focus with each calculation.

There are two main approaches to reducing cognitive load:

- break the task into smaller components
- practice the task or parts of the task until they are so easily retrieved from long-term memory that they are essentially automatic.

A lot of editing goes on!

Memories aren't made right away.

When you acquire new information, you give it its first editing, by:

- the aspects you pay attention to
- the connections you make
- the way you think about it.

Your unconscious mind then continues to edit the information for several hours after, as it stabilizes the new memory codes and tries to fit them into your existing network.

It then works on them even more as it 'consolidates' them into their final forms, as you sleep. During this process, some bits are dropped, other bits emphasized; connections may be made stronger, or cut entirely.

Then, every time you retrieve the memory codes, you open them up for a little more tweaking.

Memory is not fixed. Memory is not an end-product.

Memory is an ongoing, and never-ending, process.

And more than any other type of memory, this is what learning is: editing, re-editing, and re-re-editing the information we hold in memory.

Moreover, while new memories are being stabilized and consolidated, they are vulnerable to interference. Other new memories will compete with them, and may affect them.

Bottom line: Learning is not simply something that happens at the time you receive the information! What happens in the next twenty-four hours is critical to your success.

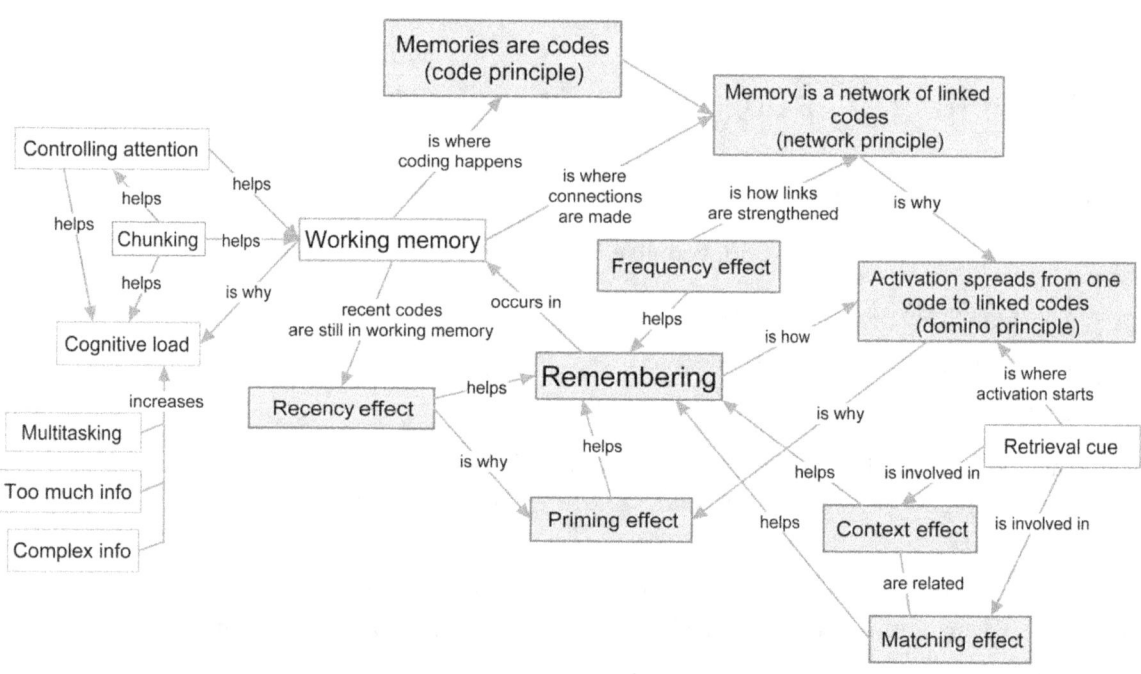

Review Questions

1. How easily you remember something depends on:

    a. How good your memory is

    b. How well you encoded the information

    c. How closely the memory code matches what you're looking for

    d. The retrieval cue you started with

    e. How important the information is

2. Encoding is about:

    a. Picking out selected information to place in long-term memory

    b. Connecting new information to information in the long-term memory network

    c. Transforming information into an edited code unique to you

    d. Turning information into a pattern of neural activity

3. What basic memory principle does the strategy of free association rely on?

    a. Code principle

    b. Network principle

    c. Domino principle

    d. Recency effect

4. If you think of your brain as a library, then working memory is most comparable to:

    a. A book on the shelf

    b. The book in your hand

    c. The page you are looking at

    d. The words you are reading

5. The 'cognitive load' of a memory task or situation refers to:

   a. How difficult it is

   b. How much time and effort you need to process it

   c. How much demand it puts on your working memory

   d. How much information needs to be processed

   e. How much information needs to be ignored

Approaching your Learning

However impatient you might be to get straight into your subject, too much haste is counter-productive. Work smarter by approaching your learning thoughtfully: think about what you want to achieve. What do you already know, and what do you need to learn? How hard is the material, and what strategies would be most appropriate?

Approaching your learning involves setting goals, evaluating the material and situation, and getting into the right mind-set.

One common problem students have is leaping straight into taking notes. But unless the material is very simple (in which case your notes should be minimal), you're going to need to take a few bites at any text. ('Text' is used throughout this book to refer to any written material, not simply what you might read in a traditional textbook.)

First bite is **priming** and establishing your goals.

Second bite is getting a preliminary 'big picture' understanding of the text.

Third bite is reading the text.

Fourth bite is to begin your note-taking.

1. Priming
2. Big picture
3. Reading
4. Note-taking

Priming

Your memory is a vast network of networks — a universe of some 100 billion neurons, a potential 5 million trillion connections.

This is a filing system of unimaginable proportions.

So how do we find anything?

Organization is crucial, obviously. But the first step in organizing well is to get the right network ready and waiting.

In psychological terms, this is called priming.

In study terms, this means:

> Before going into a lecture:
>> ★ do any reading suggested, or
>> ★ look back over your previous lecture's notes, or
>> ★ simply think about what you expect the lecture to be about.
> Before reading your textbook:
>> ★ read the table of contents, and (if available) the introduction.
> Before tackling a particular section in the textbook:
>> ★ read any summaries, such as an **overview** or an **advance organizer**, or end-of-chapter summary

- ★ skim the **headings**
- ★ study the images.
- ➢ Before any lecture or reading, work out what you want from the text or lecture.

Goal setting

Putting your goal into specific words is vital if you want to get the most out of your learning material.

Knowing what you want out of the material enables you to:
- ➢ direct your attention wisely
- ➢ select the important information (important to you, that is)
- ➢ minimize the time you waste
- ➢ choose the best strategies for learning the material.

If you have no specific questions that need answering, then you need to come up with some! That's the other major purpose of priming. Use your preview of the material to work out what you need to know — the more precise you can be, the better.

In the beginning, or with difficult material, you may find you need to explore the material in greater depth before you can work out your goals. That's fine. You'll get better with practice.

A good goal includes:
- ➢ the information you want, *and*
- ➢ the situations you're learning it for: the **retrieval context**.

In study, the retrieval context includes such situations as:
- ➢ writing an essay
- ➢ writing a summary
- ➢ preparing for group discussion
- ➢ preparing for class

- preparing for an exam:
  - ★ for multi-choice questions
  - ★ for short answers
  - ★ for essay answers.

Different retrieval contexts make different demands on your understanding and your memory, and so require different strategies.

Broadly, we can match these situations with the following memory tasks:

Situation	Tasks
Preparing for an exam:	
For multi-choice questions	Recognition
For short answers	Cued recall (in response to prompts)
For essay answers	Free recall (recall in the absence of prompts)
Writing an essay	Organization; comprehension
Writing a summary	Selection
Preparing for group discussion	Priming; cued recall
Preparing for class	Priming

Evaluating the text

Preparing for learning also includes evaluating the demands of the material.

For text, it's helpful to classify it at one of three different levels, according to its structure and density:

1. **simple** (straightforward text with clear connections)

2. **complex** (characterized by many changes of topic and more than one level of information)
3. **difficult** (dense text with many topic changes, often unclear, inconsistent and/or abstract).

These different types of text require progressively more sophisticated strategies.

To assess the difficulty of text:

- assess density:

    How many different ideas are there in each paragraph? How many on a page?

- assess the effectiveness of the structure:

    Is it divided into logical sections? Do the headings encapsulate the themes of the sections? Are changes of theme signaled by headings?

- look for the presence of effective cues:

    Are key points signaled in some way (by headings; **highlighting**; use of clear signal words)?

- assess connectivity:

    Is the information in each section meaningfully connected? How many changes of topic are there?

- assess complexity:

    Can important concepts be easily conveyed in single words or brief phrases?

- assess style:

    Is it formal or informal? A text written in a casual, chatty style is more easily read and understood than a dry, academic one.

You may find it helpful initially to use the simple assessment tool shown on the next page to help you sum these factors up.

Using the sample texts in the following pages, let's see this in practice. These texts will be used as examples throughout the book.

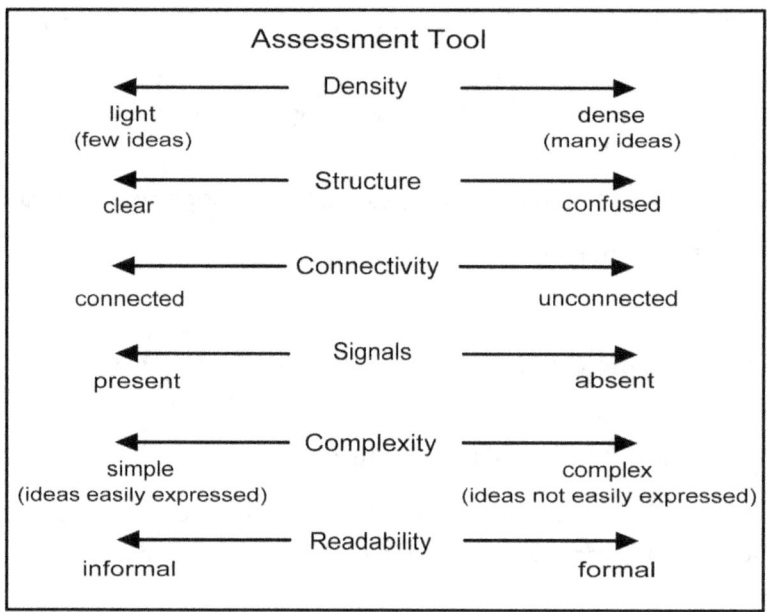

How blood flows

We all know that blood flows through our body in continuous motion, and that our heart is the pump that drives this motion. But the circulatory system is best understood not as a single system but in terms of its three constituent parts — pulmonary circulation (involving the lungs), coronary circulation (involving the heart), and systemic circulation (involving the blood vessels).

Pulmonary circulation is the movement of blood from the heart to the lungs, an d to the heart again.

The heart has four chambers — the upper chambers are called atriums; the lower chambers are called ventricles. Blood, with all the waste products it has collected in its journey through the body (most particularly carbon dioxide), enters the heart through the right atrium, via two large veins — the inferior vena cava and the superior vena cava. The inferior vena cava carries the blood from the lower half of the body; the superior from the upper half.

When the right atrium is filled with blood, it contracts, pushing the blood into the right ventricle, which then likewise contracts, pushing the blood into the pulmonary artery. The pulmonary artery carries the blood to the lungs, where carbon dioxide and oxygen are exchanged. The blood, now cleaned of its waste and rich in oxygen (because the oxygen drawn into the lungs through breathing binds with blood), is then carried by the pulmonary veins back to the heart — to the left atrium this time. From whence, in the same process as before, it passes through to the left ventricle, and then leaves the heart through the main artery — the biggest artery in the body — the aorta. From there, it begins to circulate throughout the body.

Coronary circulation refers to the movement of blood *within* the heart. Heart tissue needs nourishment, and this nourishment comes through capillaries in the heart.

So the heart and lungs have their own systems; systemic circulation is the part of the circulatory system that supplies nourishment to the tissues throughout the rest of the body. It does this, of course, through the blood vessels —arteries, veins, and capillaries. Arteries carry blood away *from* the heart. Veins carry blood to the heart. Capillaries connect the arteries to veins.

Because the heart is a pump, blood comes out in spurts, causing the outflow to vary in volume and speed. This means that blood flow can occur at a high pressure, which is why the arteries need to have thick walls, and why they need to be able to expand and contract to accommodate the changes in pressure.

Veins, on the other hand, carry blood to the heart in a continuous, even flow, and are therefore thinner than arteries, and less elastic.

Systemic circulation begins with the aorta. The aorta branches into many smaller arteries (the smallest are called arterioles), which carry the fresh, oxygenated blood through the body. Finally, the blood reaches the capillaries, where the oxygen and nutrients carried by the blood are released. The de-oxygenated blood now enters the veins, to travel back to the heart and begin its journey once more.

A simple text.

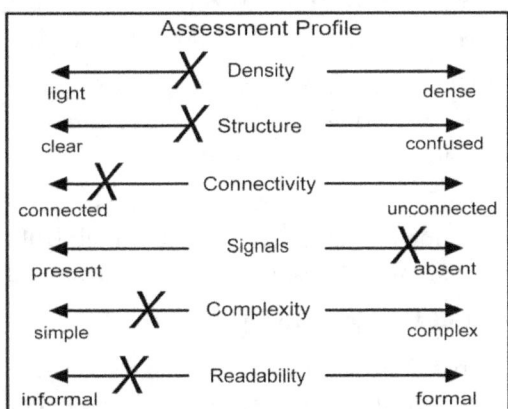

Early America

The First Americans

At the height of the Ice Age, between 34,000 and 30,000 B.C., much of the world's water was locked up in vast continental ice sheets. As a result, the Bering Sea was hundreds of meters below its current level, and a land bridge, known as Beringia, emerged between Asia and North America. At its peak, Beringia is thought to have been some 1,500 kilometers wide. A moist and treeless tundra, it was covered with grasses and plant life, attracting the large animals that early humans hunted for their survival.

The first people to reach North America almost certainly did so without knowing they had crossed into a new continent. They would have been following game, as their ancestors had for thousands of years, along the Siberian coast and then across the land bridge.

Once in Alaska, it would take these first North Americans thousands of years more to work their way through the openings in great glaciers south to what is now the United States. Evidence of early life in North America continues to be found. Little of it, however, can be reliably dated before 12,000 B.C.; a recent discovery of a hunting lookout in northern Alaska, for example, may date from almost that time. So too may the finely crafted spear points and items found near Clovis, Ne Mexico.

Similar artifacts have been found at sites throughout North and South America, indicating that life was probably already well established in much of the Western Hemisphere by some time prior to 10,000 B.C. Around that time the mammoth began to die out and the bison took its place as a principal source of food and hides for these early North Americans. Over time, as more and more species of large game vanished whether from overhunting or natural causes plants, berries, and seeds became an increasingly important part of the early American diet. Gradually, foraging and the first attempts at primitive agriculture appeared. Native Americans in what is now central Mexico led the way, cultivating corn, squash, and beans, perhaps as early as 8,000 B.C. Slowly, this knowledge spread northward.

By 3,000 B.C., a primitive type of corn was being grown in the river valleys of New Mexico and Arizona. Then the first signs of irrigation began to appear, and, by 300 B.C., signs of early village life.

By the first centuries A.D., the Hohokam were living in settlements near what is now Phoenix, Arizona, where they built ball courts and pyramid like mounds reminiscent of those found in Mexico, as well as a canal and irrigation system.

Mound Builders And Pueblos

The first Native-American group to build mounds in what is now the United States often are called the Adenans. They began constructing earthen burial sites and fortifications around 600 B.C. Some mounds from that era are in the shape of birds or serpents; they probably served religious purposes not yet fully understood.

The Adenans appear to have been absorbed or displaced by various groups collectively known as Hopewellians. One of the most important centers of their culture was found in southern Ohio, where the remains of several thousand of these mounds still can be seen. Believed to be great traders, the Hopewellians used and exchanged tools and materials across a wide region of hundreds of kilometers.

By around 500 A.D., the Hopewellians disappeared, too, gradually giving way to a broad group of tribes generally known as the Mississippians or Temple Mound culture. One city, Cahokia, near Collinsville, Illinois, is thought to have had a population of about 20,000 at its peak in the early 12th century. At the center of the city stood a huge earthen mound, flattened at the top, that was 30 meters high and 37 hectares at the base. Eighty other mounds have been found nearby.

Cities such as Cahokia depended on a combination of hunting, foraging, trading, and agriculture for their food and supplies. Influenced by the thriving societies to the south, they evolved into complex hierarchical societies that took slaves and practiced human sacrifice.

In what is now the southwest United States, the Anasazi, ancestors of the modern Hopi Indians, began building stone and adobe pueblos around the year 900. These unique and amazing apartment-like structures were often built along cliff faces; the most famous, the "cliff palace" of Mesa Verde, Colorado, had more than 200 rooms. Another site, the Pueblo Bonito ruins along New Mexico's Chaco River, once contained more than 800 rooms.

Perhaps the most affluent of the pre-Columbian Native Americans lived in the Pacific Northwest, where the natural abundance of fish and raw materials made food supplies plentiful and permanent villages possible as early as 1,000 B.C. The opulence of their "potlatch" gatherings remains a standard for extravagance and festivity probably unmatched in early American history.

Native-American Cultures

The America that greeted the first Europeans was, thus, far from an empty wilderness. It is now thought that as many people lived in the Western Hemisphere as in Western Europe at that time — about 40 million. Estimates of the number of Native Americans living in what is now the United States at the onset of European colonization range from two to 18 million, with most historians tending toward the lower figure. What is certain is the devastating effect that European disease had on the indigenous population practically from the time of initial contact. Smallpox, in particular, ravaged whole communities and is thought to have been a much more direct cause of the precipitous decline in the Indian population in the 1600s than the numerous wars and skirmishes with European settlers.

Indian customs and culture at the time were extraordinarily diverse, as could be expected, given the expanse of the land and the many different environments to which they had adapted. Some generalizations, however, are possible. Most tribes, particularly in the wooded eastern region and the Midwest, combined aspects of hunting, gathering, and the cultivation of maize and other products for their food supplies. In many cases, the women were responsible for farming and the distribution of food, while the men hunted and participated in war.

By all accounts, Native-American society in North America was closely tied to the land. Identification with nature and the elements was integral to religious beliefs. Their life was essentially clan-oriented and communal, with children allowed more freedom and tolerance than was the European custom of the day.

Although some North American tribes developed a type of hieroglyphics to preserve certain texts, Native-American culture was primarily oral, with a high value placed on the recounting of tales and dreams. Clearly, there was a good deal of trade among various groups and strong evidence exists that neighboring tribes maintained extensive and formal relations — both friendly and hostile.

The First Europeans

The first Europeans to arrive in North America — at least the first for whom there is solid evidence — were Norse, traveling west from Greenland, where Erik the Red had founded a settlement around the year 985. In 1001 his son Leif is thought to have explored the northeast coast of what is now Canada and spent at least one winter there.

While Norse sagas suggest that Viking sailors explored the Atlantic coast of North America down as far as the Bahamas, such claims remain unproven. In 1963, however, the ruins of some Norse houses dating from that era were discovered at L'Anse-aux-Meadows in northern Newfoundland, thus supporting at least some of the saga claims.

In 1497, just five years after Christopher Columbus landed in the Caribbean looking for a western route to Asia, a Venetian sailor named John Cabot arrived in Newfoundland on a mission for the British king. Although quickly forgotten, Cabot's journey was later to provide the basis for British claims to North America. It also opened the way to the rich fishing grounds off George's Banks, to which European fishermen, particularly the Portuguese, were soon making regular visits.

Columbus never saw the mainland of the future United States, but the first explorations of it were launched from the Spanish possessions that he helped establish. The first of these took place in 1513 when a group of men under Juan Ponce de León landed on the Florida coast near the present city of St. Augustine.

With the conquest of Mexico in 1522, the Spanish further solidified their position in the Western Hemisphere. The ensuing discoveries added to Europe's knowledge of what was now named America — after the Italian Amerigo Vespucci, who wrote a widely popular account of his voyages to a "New World". By 1529 reliable maps of the Atlantic coastline from Labrador to Tierra del Fuego had been drawn up, although it would take more than another century before hope of discovering a "Northwest Passage" to Asia would be completely abandoned.

Among the most significant early Spanish explorations was that of Hernando De Soto, a veteran conquistador who had accompanied Francisco Pizarro in the conquest of Peru. Leaving Havana in 1539, De Soto's expedition landed in Florida and ranged through the southeastern United States as far as the Mississippi River in search of riches.

Another Spaniard, Francisco Vázquez de Coronado, set out from Mexico in 1540 in search of the mythical Seven Cities of Cibola. Coronado's travels took him to the Grand Canyon and Kansas, but failed to reveal the gold or treasure his men sought. However, his party did leave the peoples of the region a remarkable, if unintended, gift: Enough of his horses escaped to transform life on the Great Plains. Within a few generations, the Plains Indians had become masters of horsemanship, greatly expanding the range and scope of their activities.

While the Spanish were pushing up from the south, the northern portion of the present-day United States was slowly being revealed through the journeys of men such as Giovanni da Verrazano. A Florentine who sailed for the French, Verrazano made landfall in North Carolina in 1524, then sailed north along the Atlantic Coast past what is now New York harbor.

A decade later, the Frenchman Jacques Cartier set sail with the hope — like the other Europeans before him — of finding a sea passage to Asia. Cartier's expeditions along the St. Lawrence River laid the foundation for the French claims to North America, which were to last until 1763.

Following the collapse of their first Quebec colony in the 1540s, French Huguenots attempted to settle the northern coast of Florida two decades later. The Spanish, viewing the French as a threat to their trade route along the Gulf Stream, destroyed the colony in 1565. Ironically, the leader of the Spanish forces, Pedro Menéndez, would soon establish a town not far away — St. Augustine. It was the first permanent European settlement in what would become the United States.

A simple text.

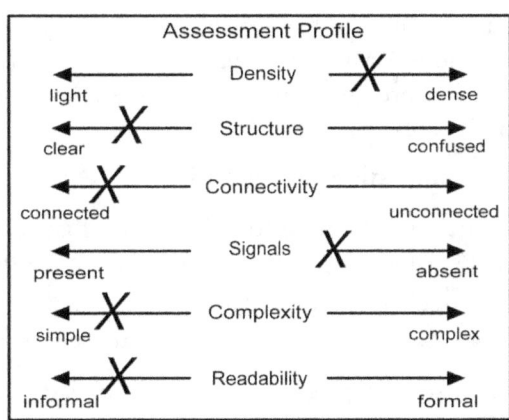

Approaching your Learning

The role of consolidation in memory

'Consolidation' is a term that is bandied about a lot in recent memory research. Here's my take on what it means.

How information becomes a memory

Initially, information is thought to be encoded as patterns of neural activity - cells 'talking' to each other. Later, the information is coded in more persistent molecular or structural formats (e.g., the formation of new synapses). It has been assumed that once this occurs, the memory is 'fixed' — a permanent, unchanging, representation.

With new techniques, it has indeed become possible to observe these changes. Researchers found that the changes to a cell that occurred in response to an initial stimulation lasted some three to five minutes and disappeared within five to 10 minutes. If the cell was stimulated four times over the course of an hour, however, the synapse would actually split and new synapses would form, producing a (presumably) permanent change.

Memory consolidation theory

The hypothesis that new memories consolidate slowly over time was proposed 100 years ago, and continues to guide memory research. In modern consolidation theory, it is assumed that new memories are initially "labile" and sensitive to disruption before undergoing a series of processes (e.g., glutamate release, protein synthesis, neural growth and rearrangement) that render the memory representations progressively more stable. It is these processes that are generally referred to as "consolidation".

Recently, however, the idea has been gaining support that stable representations can revert to a labile state on reactivation.

Memory as reconstruction

In a way, this is not surprising. We already have ample evidence that retrieval is a dynamic process during which new information merges with and modifies the existing representation — memory is now seen as reconstructive, rather than a simple replaying of stored information

Reconsolidation of memories

Researchers who have found evidence that supposedly stable representations have become labile again after reactivation, have called the process "reconsolidation", and suggest that consolidation, rather than being a one-time event, occurs repeatedly every time the representation is activated.

This raises the question: does reconsolidation involve *replacing* the previously stable representation, or the establishment of a new representation, that coexists with the old?

Whether reconsolidation is the creating of a new representation, or the modifying of an old, is this something other than the reconstruction of memories as they are retrieved? In other words, is this recent research telling us something about consolidation (part of the encoding process), or something about reconstruction (part of the retrieval process)?

Hippocampus involved in memory consolidation

The principal player in memory consolidation research, in terms of brain regions, is the hippocampus. The hippocampus is involved in the recognition of place and the consolidation of contextual memories, and is part of a region called the medial temporal lobe (MTL), that also includes the perirhinal, parahippocampal, and entorhinal cortices. Lesions in the medial temporal lobe typically produce amnesia characterized by the disproportionate loss of recently acquired memories. This has been interpreted as evidence for a memory consolidation process.

Some research suggests that the hippocampus may participate only in consolidation processes lasting a few years. The entorhinal cortex, on the other hand, gives evidence of temporally graded changes extending up to 20 years, suggesting that it is this region that participates in memory consolidation over decades. The entorhinal cortex is damaged in the early stages of Alzheimer's disease.

There is, however, some evidence that the hippocampus can be involved in older memories — perhaps when they are particularly vivid.

A recent idea that has been floated suggests that the entorhinal cortex, through which all information passes on its way to the hippocampus, handles "incremental learning" — learning that requires repeated experiences. "Episodic learning" — memories that are stored after only one occurrence — might be mainly stored in the hippocampus.

This may help explain the persistence of some vivid memories in the hippocampus. Memories of emotionally arousing events tend to be more vivid and to persist longer than do memories of neutral or trivial events, and are, moreover, more likely to require only a single experience.

Whether or not the hippocampus may retain some older memories, the evidence that some memories might be held in the hippocampus for several years, only to move on, as it were, to another region, is another challenge to a simple consolidation theory.

Memory less stable than we thought

So where does all this leave us? What *is* consolidation? *Do* memories reach a fixed state?

My own feeling is that, no, memories don't reach this fabled 'cast in stone' state. Memories are subject to change every time they are activated (such activation doesn't have to bring the memory to your conscious awareness). But consolidation traditionally (and logically) refers to encoding processes. It is reasonable, and useful, to distinguish between:

- the initial encoding, the 'working memory' state, when new information is held precariously in shifting patterns of neural activity,

- the later encoding processes, when the information is consolidated into a more permanent form with the growth of new connections between nerve cells,

- the (possibly much) later retrieval processes, when the information is retrieved in, most probably, a new context, and is activated anew

I think that 'reconsolidation' is a retrieval process rather than part of the encoding processes, but of course, if you admit retrieval as involving a return to the active state and a modification of the original representation in line with new associations, then the differences between retrieval and encoding become less evident.

When you add to this the possibility that memories might 'move' from one area of the brain to another after a certain period of time (although it is likely that the triggering factor is not time per se), then you cast into disarray the whole concept of memories becoming stable.

Perhaps our best approach is to see memory as a series of processes, and consolidation as an agreed-upon (and possibly arbitrary) subset of those processes.

A simple/complex text (i.e., at the simple end of complex).

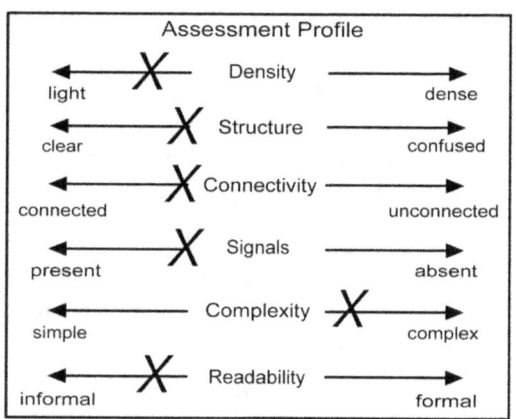

The Relationship Of Ozone And Ultraviolet Radiation: Why Is Ozone So Important?

In this section, we will explore what is ozone and what is ultraviolet radiation. We then will explore the relationship between ozone and ultraviolet radiation from the sun. It is here that ozone plays its essential role in shielding the surface from harmful ultraviolet radiation. By screening out genetically destructive ultraviolet radiation from the Sun, ozone protects life on the surface of Earth. It is for this reason that ozone acquires an enormous importance. It is why we study it so extensively.

About 90% of the ozone in our atmosphere is contained in the stratosphere, the region from about 10 to 50-km (32,000 to 164,000 feet) above Earth's surface. Ten percent of the ozone is contained in the troposphere, the lowest part of our atmosphere where all of our weather takes place. Measurements taken from instruments on the ground, flown on balloons, and operating in space show that ozone concentrations are greatest betwee about 15 and 30 km.

Although ozone concentrations are very small, typically only a few molecules O^3 per million molecules of air, these ozone molecules are vitally important to life because they absorb the biologically harmful ultraviolet radiation from the Sun. There are three different types of ultraviolet (UV) radiation, based on the wavelength of the radiation. These are referred to as UV-a, UV-b, and UV-c. UV-c (red) is entirely screened out by ozone around 35 km altitude, while most UV-a (blue) reaches the surface, but it is not as genetically damaging, so we don't worry about it too much. It is the UV-b (green) radiation that can cause sunburn and that can also cause genetic damage, resulting in things like skin cancer, if exposure to it is prolonged. Ozone screens out most UV-b, but some reaches the surface. Were the ozone layer to decrease, more UV-b radiation would reach the surface, causing increased genetic damage to living things.

Because most of the ozone in our atmosphere is contained in the stratosphere, we refer to this region as the stratospheric ozone layer. In contrast to beneficial stratospheric ozone, tropospheric ozone is a pollutant found in high concentrations in smog. Though it too absorbs UV radiation, breathing it in high levels is unhealthy, even toxic. The high reactivity of ozone results in damage to the living tissue of plants and animals. This damage by heavy tropospheric ozone pollution is often manifested as eye and lung irritation. Tropospheric ozone is mainly produced during the daytime in polluted regions such as urban areas. Significant government efforts are underway to regulate the gases and emissions that lead to this harmful pollution, and smog alerts are regular occurrences in polluted urban areas.

To appreciate the importance of stratospheric ozone, we need to understand something of the Sun's output and how it impacts living systems. The Sun produces radiation at many different wavelengths. These are part of what is known as the electromagnetic (EM) spectrum. EM radiation includes everything from radio waves (very long wavelengths) to X-rays and gamma rays (very tiny wavelengths). EM radiation is classified by wavelength, which is a measure of how energetic is the radiation. The energy of a tiny piece or "packet" of radiation (which we call a photon) is inversely proportional to its wavelength.

The human eye can detect wavelengths in the region of the spectrum from about 400 nm (nanometers or billionths of a meter) to about 700 nm. Not surprisingly, this is called the visible region of the spectrum. All the colors of light (red, orange, yellow, green, blue, and violet) fall inside a small wavelength band. Whereas radio waves have wavelengths on the order of meters, visible light waves have wavelengths on the order of billionths of a meter. Such a tiny unit is called a nanometer (1 nm= 10^{-9} m). At one end of the visible "color" spectrum is red light. Red light has a wavelength of about 630 nm. Near the opposite end of the color spectrum is blue light, and at the very opposite end is violet light. Blue light has a wavelength of about 430 nm. Violet light has a wavelength of about 410 nm. Therefore, blue light is more energetic than red light because of its shorter wavelength, but it is less energetic than violet light, which has an even shorter wavelength. Radiation with wavelengths shorter than those of violet light is called ultraviolet radiation.

The Sun produces radiation that is mainly in the visible part of the electromagnetic spectrum. However, the Sun also generates radiation in ultraviolet (UV) part of the spectrum. UV wavelengths range from 1 to 400 nm. We are concerned about ultraviolet radiation because these rays are energetic enough to break the bonds of DNA molecules (the molecular carriers of our genetic coding), and thereby damage cells. While most plants and animals are able to either repair or destroy damaged cells, on occasion, these damaged DNA molecules are not repaired, and can replicate, leading to dangerous forms of skin cancer (basal, squamous, and melanoma).

Solar flux refers to the amount of solar energy in watts falling perpendicularly on a surface one square centimeter, and the units are watts per cm^2 per nm. Because of the strong absorption of UV radiation by ozone in the stratosphere, the intensity decreases at lower altitudes in the atmosphere. In addition, while the energy of an individual photon is greater if it has a shorter wavelength, there are fewer photons at the shorter wavelengths, so the Sun's total energy output is less at the shorter wavelengths. Because of ozone, it is virtually impossible for solar ultraviolet to penetrate to Earth's surface. For radiation with a wavelength of 290 nm, the intensity at Earth's surface is 350 million times weaker than at the top of the atmosphere. If our eyes detected light at less than 290 nm instead of in the visible range, the world would be very dark because of the ozone absorption!

To appreciate how important this ultraviolet radiation screening is, we can consider a characteristic of radiation damage called an action spectrum. An action spectrum gives us a measure of the relative effectiveness of radiation in generating a certain biological response over a range of wavelengths. This response might be erythema (sunburn), changes in plant growth, or changes in molecular DNA. Fortunately, where DNA is easily damaged (where there is a high probability), ozone strongly absorbs UV. At the longer wavelengths where ozone absorbs weakly, DNA damage is less likely. If there was a 10% decrease in ozone, the amount of DNA damaging UV would increase by about 22%. Considering that DNA damage can lead to maladies like skin cancer, it is clear that this absorption of the Sun's ultraviolet radiation by ozone is critical for our well-being.

While most of the ultraviolet radiation is absorbed by ozone, some does make it to Earth's surface. Typically, we classify ultraviolet radiation into three parts, UV-a (320-400 nm), UV-b (280-320 nm), and UV-c (200-280 nm). Sunscreens have been developed by commercial manufacturers to protect human skin from UV radiation. The labels of these sunscreens usually note that they screen both UV-a and UV-b. Why not also screen for UV-c radiation? When UV-c encounters ozone in the mid-stratosphere, it is quickly absorbed so that none reaches Earth's surface. UV-b is partially absorbed and UV-a is barely absorbed by ozone. Ozone is so effective at absorbing the extremely harmful UV-c that sunscreen manufacturers don't need to worry about UV-c. Manufacturers only need to eliminate skin absorption of damaging UV-b and less damaging UV-a radiation.

The screening of ultraviolet radiation by ozone depends on other factors, such as time of day and season. The angle of the Sun in the sky has a large effect on the UV radiation. When the Sun is directly overhead, the UV radiation comes straight down through our atmosphere and is only absorbed by overhead ozone. When the Sun is just slightly above the horizon at dawn and dusk, the UV radiation must pass through the atmosphere at an angle. Because the UV passes through a longer distance in the atmosphere, it encounters more ozone molecules and there is greater absorption and, consequently, less UV radiation striking the surface.

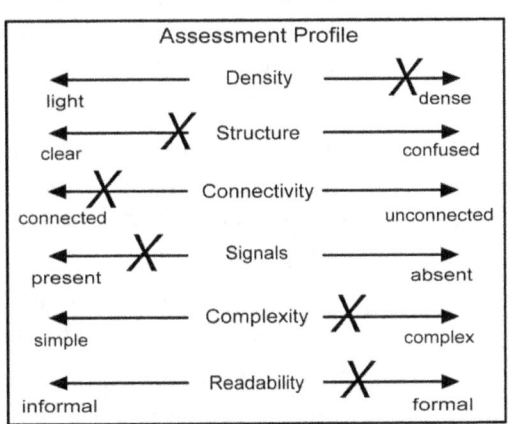

A complex text.

(text adapted from NASA's Stratospheric Ozone Electronic Textbook, http://www.ccpo.odu.edu/SEES/ozone/oz_class.htm)

Introducing brain cells

The brain contains two types of nerve cell: *neurons* and *glia*. There are, roughly, some 100 billion neurons in the human brain. There are ten times as many glia. Yet it is neurons we talk about all the time. Of course, it is neurons that are important! But glia are the "glue" that hold the neurons together, and the latest research suggests that glia are more important than we have thought.

But neurons, although only 10% of our brain cells, do perform most of the information processing, which is why they have always been the principal focus of research.

Neurons, like glia, are a broad class of cells. There are several different types of neuron, but they all have certain attributes in common. They all have a *cell body / soma*, containing the *cell nucleus*. They all have thin tubes radiating from the cell body. These are called *neurites*, and come in two flavors: *axons* and *dendrites*. There is usually only one axon, and it is of uniform thickness, and very long by comparison with dendrites (axons can be over a meter in length). A neuron contains many dendrites, which are very short (rarely more than 2mm) and usually taper to a fine point.

It is the axon that carries the output of the neuron. It is the dendrites, which come in contact with many axons, that receive the incoming signals.

The soma, the cell body, is roughly spherical and contains the same organelles contained in any animal cell. The most important are the nucleus, the *rough endoplasmic reticulum*, the *smooth endoplasmic reticulum*, the *Golgi apparatus*, and the *mitochondria*.

The nucleus holds the chromosomes, which contain the DNA.

Endoplasmic reticulum are, very basically, folded membranes. Rough endoplasmic reticulum exist in all cells, but are particularly abundant in neurons. Rough ER contains *ribosomes*, tiny balls vitally involved in protein synthesis. Rough ER contrasts with smooth ER, which is just the same except that the membranes don't contain ribosomes. The function of smooth ER depends on its location within the cell. Most smooth ER plays no role in protein synthesis.

Yet another type of folded membrane is the Golgi apparatus, where processing of the proteins, after their assembly, takes place. It is thought that, among other functions, the Golgi apparatus is involved in sorting proteins for delivery to different parts of the neuron.

The last vital structure within the neuron is the mitochondrion. The function of this type of cell is to supply the energy the cell needs to function.

The shape of a neuron is governed by its *cytoskeleton*. The cytoskeleton consists of three types of element: *microtubules*, *microfilaments*, and *neurofilaments*.

Of these, by far the biggest are the microtubules, which may be thought of as hollow tubes that run through neurites. Neurofilaments are between microtubules and microfilaments in size. Similar filaments are found in cells other than neurons — one such is keratin, which, bundled together, makes hair. Unlike microtubules and microfilaments, both of which are made up of polymers, neurofilaments are made from single long protein molecules. This makes them very strong, and also very stable. However, neurofilaments can cause problems — the neurofibrillary tangles characteristic of Alzheimer's disease are neurofilaments gone wild.

Microfilaments are only about as thick as the cell membrane. Although they're found throughout the neuron, they're particularly abundant in the neurites.

These structures — the soma, the organelles within, the membrane, the cytoskeleton — exist in all cells, but now we come to a part of the neuron that is unique to neurons: the axon. As was mentioned, the axon is the means by which the neuron can send its message on — it's the output mechanism. The important thing to remember is that neurons, unlike other cells in the body, aren't in physical contact with each other. To communicate with each other they need something to leap the gap between them.

The synapse is the point of contact between the neurons, and information flows from across the gap between neurons in a process called *synaptic transmission*, which involves the release of chemicals called *neurotransmitters*.

The other type of neurite is the dendrites, which derive their name from the Greek for tree. This is because the dendrites, as they branch out from the soma in profusion, resemble the branches of a tree.

Neurons can be categorized in various ways. They can be classified by the number of neurites they have. A neuron with one neurite is *unipolar*; one with two is *bipolar*; one with more is *multipolar*. Most neurons are multipolar.

They can also be classified according to the shape and size of the dendritic tree, (e.g., pyramidal cells, stellate cells), or according to whether their dendrites have spines (*spiny* vs *aspinous*). They can also be classified according to the connections they make: *primary sensory neurons* connect with sensory surfaces; *motor neurons* connect with muscles; *interneurons* connect with other neurons. Most neurons are interneurons.

Neurons can also be classified according to axon length: *projection neurons* (or *Golgi Type I*) have long axons that extend from one part of the brain to another; *local circuit neurons* (or *Golgi Type II*) have short axons.

And finally, they can be classified according to chemistry, that is, on the particular neurotransmitters they release.

But most of the brain is taken up by glia, the support cells. Most glia are a type

Approaching your Learning

called *astrocytes*. It's the astrocytes that fill the spaces between neurons. *Oligodendroglia* provide the insulation for axons in the brain and spinal cord. *Schwann cells* are similar to the oligodendroglia, fulfilling the same function outside the brain and spinal cord, in the peripheral nervous system.

A complex/difficult text (i.e., at the difficult end of complex).

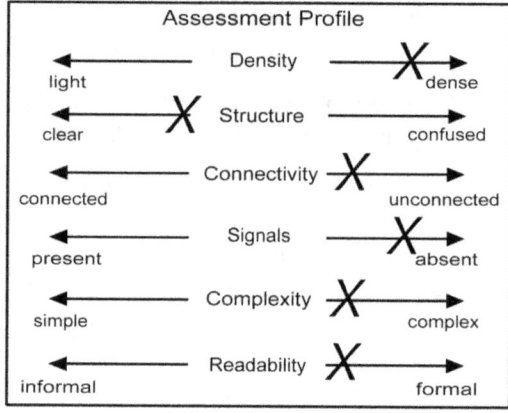

Human Gene Affects Memory

NIH scientists have shown that a common gene variant influences memory for events in humans by altering a growth factor in the brain's memory hub. On average, people with a particular version of the gene that codes for brain derived neurotrophic factor (BDNF) performed worse on tests of episodic memory — tasks like recalling what happened yesterday. They also showed differences in activation of the hippocampus, a brain area known to mediate memory, and signs of decreased neuronal health and interconnections. These effects are likely traceable to limited movement and secretion of BDNF within cells, according to the study, which reveals how a gene affects the normal range of human memory, and confirms that BDNF affects human hippocampal function much as it does animals'.

Long known to be critical for the growth and survival of neurons, BDNF has also recently been shown to play a key role in memory and hippocampal function in animals. To find out if it works similarly in humans, the researchers explored the consequences of a tiny variance in the human BDNF gene, where its molecular makeup differs slightly across individuals.

People inherit two copies of the BDNF gene — one from each parent — in either of two versions. Slightly more than a third inherit at least one copy of a version nicknamed "met", which the researchers have now linked to poorer memory. It's called "met" because its chemical sequence contains the amino acid methionine in a location where the more common version, "val", contains valine.

"We are finding that this one amino acid substitution exerts a substantial influence on human memory, presumably because of its effects on the biology of the hippocampus," said Weinberger.

Despite its negative effect on memory, the "met" version's survival in the human genome suggests that it "may confer some compensatory advantage in other biological processes", note the researchers. Although they found that it does not confer increased susceptibility to schizophrenia, they suggest that the "met" variant might contribute to risk for — or increase functional impairment in — other disorders involving hippocampal dysfunction, such as Alzheimer's disease or mood disorders.

Drawing on participants in the NIMH intramural sibling study of schizophrenia, Egan and colleagues first assessed their hippocampal function and related it to their BDNF gene types.

Among 641 normal controls, schizophrenia patients, and their unaffected siblings, those who had inherited two copies of the "met" variant scored significantly lower than their matched peers on tests of verbal episodic (event) memory. Most notably, normal controls with two copies of "met" scored 40 percent on delayed recall, compared to 70 percent for those with two copies of "val". BDNF gene type had no significant effect on tests of other types of memory, such as semantic or working memory.

The researchers then measured brain activity in two separate groups of healthy subjects while they were performing a working memory task that normally turns off hippocampus activity. Functional magnetic resonance imaging (fMRI) scans revealed that those with one copy of "met" showed a pattern of activation along the sides of the hippocampus, in contrast to lack of activation among those with two copies of "val".

Next, an MRI scanner was used to measure levels of a marker inside neurons indicating the cell's health and abundance of synapses — tiny junctions through which neurons communicate with each other. Again, subjects with one copy of "met" had lower levels of the marker, N-acetyl-aspartate (NAA), than matched individuals with two copies of "val". Analysis showed that NAA levels dropped as the number of inherited "met" variants increased, suggesting a possible "dose effect".

Unlike other growth factors, hippocampal BDNF is secreted, in part, in response to neuronal activity, making it a likely candidate for a key role in synaptic plasticity, learning and memory. To explore possible mechanisms underlying the observed "met"-related memory effect, the researchers examined the distribution, processing and secretion of the BDNF proteins expressed by the two different gene variants within hippocampal cells. When they tagged the gene variants with green fluorescent protein and introduced them into cultured neurons, they discovered that "val" BDNF spreads throughout the cell and into the branch-like dendrites that form synapses, while "met" BDNF mostly clumps inside the cell body without being transported to the synapses. To regulate memory function, BDNF must be secreted near the synapses.

"We were surprised to see that 'met' BDNF secretion can't be properly regulated by neural activity," said Lu.

The observed memory decrements are likely traceable to the failure of "met" BDNF to reach the synapses, as well as its inability to secrete in response to neuronal activity, say the researchers.

"Our study provides direct in vivo data that the molecular mechanisms related to activity dependent BDNF secretion and signaling, such as synaptic plasticity, may underlie humans' greatly expanded verbally-mediated memory system, just as it does for more rudimentary forms of memory in animals", said Egan.

In following-up their leads, the researchers are searching for a possible BDNF connection with the memory problems and hippocampal changes of Alzheimer's disease, depression and normal aging.

A difficult text.

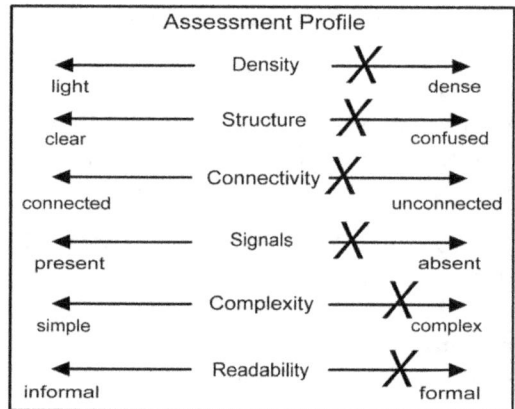

Don't feel you have to read the text to assess these attributes! While your later reading might sometimes cause you to re-evaluate, once you've become practiced at this task this initial evaluation can be done simply by scanning the structure of the text, looking for obvious signals such as headings and highlighted text (just noting their presence and number, not reading or thinking about them), and reading one or two paragraphs.

Strategies that are particularly appropriate for different levels of text difficulty

Text difficulty not only gives you an idea of how much time and effort you'll need to spend on the material (important information for goal-setting and self-monitoring), it also gives you a starting point for choosing the right strategies. This table below gives you a general idea of the strategies that are appropriate for different levels of difficulty.

Level of Difficulty	Appropriate Strategies
Simple	Highlighting; outline; topical summary
Complex	Headings; graphic organizer; multimedia summary; map; elaborative interrogation; mind map; concept map
Difficult	Self-explanation; concept map

I am not suggesting that these are the only strategies appropriate when texts are at these levels of difficulty, or that they will always be appropriate. This table is merely a guide to suggest a place to start.

Once you've

> activated your relevant knowledge (become primed),

> thought about what you want from the text (established your goals), and

> have an idea of how much time and effort you're going to need to put in (evaluated the text),

then you can dig a little deeper into the content.

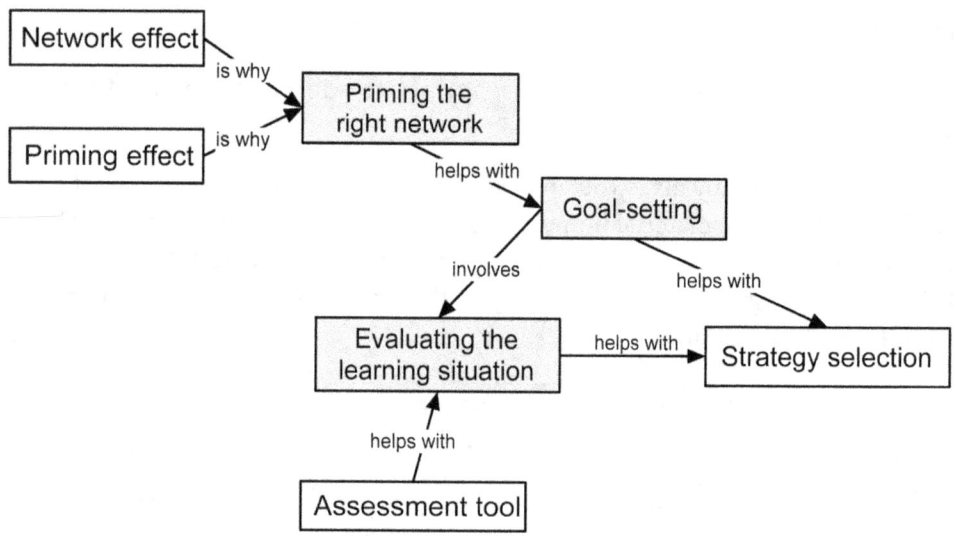

Review Questions

1. The first step when approaching a new learning situation is:

    a. Setting your goal

    b. Priming your mind

    c. Thinking about the retrieval situation

    d. Skimming the text

    e. Studying any images

2. How would you classify a text that has many topics and more than one level of information?

   a. Simple

   b. Complex

   c. Difficult

3. How would you classify a dense, abstract, and unclear text with many topics?

   a. Simple

   b. Complex

   c. Difficult

4. If you are confronted by a text that is difficult, you should:

   a. Assume you're too stupid to understand it, and give up

   b. Realize that the author is assuming a greater knowledge than you currently possess

   c. Look for appropriate strategies

   d. Take it slowly, and go through it step by step

5. Assessing the difficulty of a text helps you

   a. Set appropriate goals

   b. Choose the appropriate strategy

   c. Prime the appropriate network

   d. Monitor your learning

6. You need to read the text closely to assess the difficulty of the text.

   a. True

   b. False

7. Which of these represents the most difficult text?

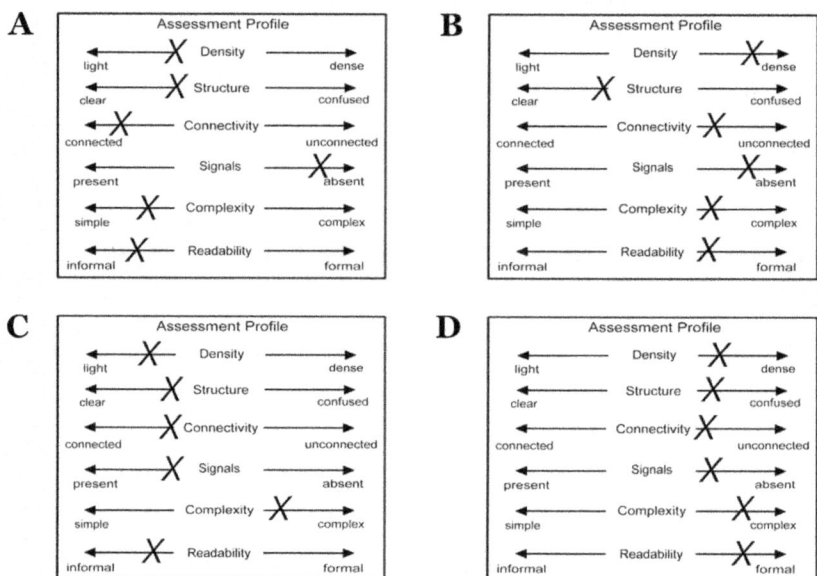

8. Which of these represents a simple text?

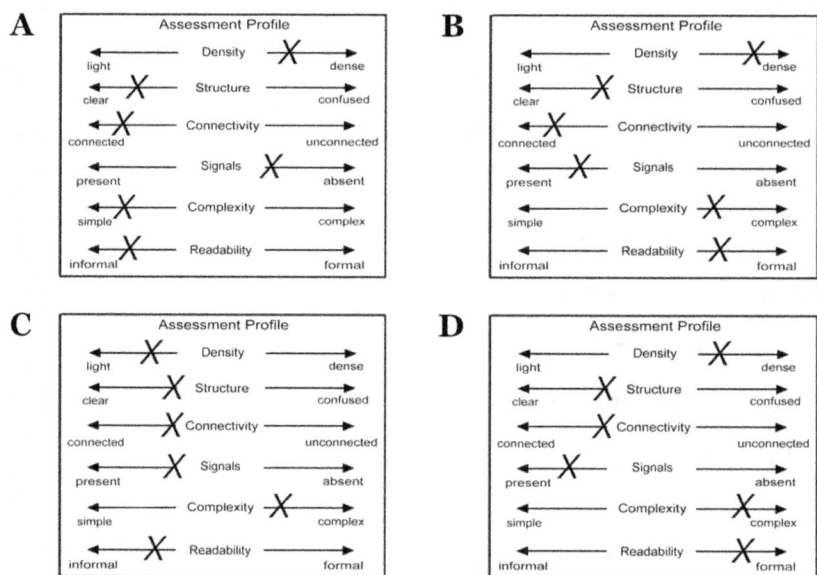

Getting an Understanding of the Text

Get a general feel for the text before reading it properly. You want to refine your ideas about what you're about to study. Use provided summaries and illustrations to help you build this preliminary understanding. You're not trying for perfect accuracy here! The important thing is to prime your brain and get yourself thinking along useful lines.

Now comes the 'big picture' step. Before properly reading the text, you want to get a general feel for what it's about and what's important. Here's an outline of the process:

> ➢ Read any summary or advance organizer first (this alerts you to what the writer thinks is important).
> ➢ Lightly skim the text: Pay attention to headings.
> ➢ Look for signal words.
> ➢ Try to identify the topic structure.
> ➢ Pay attention to illustrations.
> ➢ Pay more attention to the first and last sentences in a paragraph (compared to other sentences).

Let's look at each of these in turn.

Summaries

Types of summary

Topical summary:

> ➢ a straightforward string of factual statements; a summary of the main points
> ➢ the simplest and most common type of summary
> ➢ appears at the end of the text it summarizes.

Overview: a topical summary that appears *before* the text it summarizes.

Advance organizer:

> ➢ appears before the text
> ➢ prepares the reader for the text
> ➢ strictly speaking, is not a summary at all.

Other (less common) types of summary re-organize the text into a different format. These are more likely to appear as **illustrations** rather than summaries.

Graphic summaries include:
- ➢ **Outlines**
- ➢ **Graphic organizers**
- ➢ **Multimedia summaries**
- ➢ **Maps**

Topical summaries and overviews

Topical summaries/overviews help you learn by:
- ➢ providing a structure to help you understand the material and take effective notes
- ➢ helping you select the most important information (if it's mentioned in the summary, it's presumably of importance)
- ➢ priming your brain (prepare you for the text)
- ➢ making reading of the text faster
- ➢ making difficult or poorly organized text more easily understood
- ➢ helping you remember more topics (but perhaps at the expense of less important details).

DON'T regard summaries that are provided for you as a substitute for taking your own notes!

Why not? Because:
- ➢ much of the benefit in taking notes is about putting them into your own words, so someone else's summary isn't going to do much for you
- ➢ the writer's goals aren't necessarily (in fact, will almost certainly not be!) the same as yours.

But you can use provided summaries as a **priming tool**. To do this:
- ➢ Read the summary before reading the text.
- ➢ Reflect on it:
    - ★ call to mind any relevant knowledge you already have

- ★ think about what you expect to learn from the text
- ★ think about what you need to know (what your goal is).
- ➤ Put in your own words the essence of the summary.

Let's look at an example. Below is a short summary for the ozone text. As you read it, think about what you're reading, by:

- ➤ searching your mind for connections to familiar concepts (e.g., smog; skin cancer — neither of which are mentioned by name, but are implicit in the mention of an atmospheric pollutant close to the surface, of UV waves that damage DNA).
- ➤ noting any terms that are new to you (e.g., troposphere), so you can look for an explanation of them when you read the text.

After you've read it, imagine you're telling a friend what the text is about. What is the *heart* of this summary? Try to sum up its essence — the main topic — in one short sentence.

> Most of the ozone in our atmosphere is held in the stratosphere, in what's called the ozone layer.
>
> The ozone layer protects us from harmful ultraviolet radiation.
>
> However, some ozone is found closer to the surface, in the troposphere, where it is a pollutant.
>
> Ultraviolet waves are dangerous because they're energetic enough to damage DNA, but the shortest waves are blocked by the ozone layer, and the longest are not so damaging, so the main problem are those in the middle (UV-b).
>
> Time and season affect how much of this radiation is absorbed by the ozone layer, because the angle of the sun affects how long the radiation takes to pass through it.

Advance organizers

Advance organizers:

- ➤ appear before the text they refer to (like overviews)
- ➤ are there to prepare the student's mind (like overviews)
- ➤ include fewer details than overviews

- are written at a higher level of abstraction (*not* like overviews!)
- include information or a perspective that is not in the text (*not* like overviews)
- are about putting the text in **context**.

See how an advance organizer for the ozone text differs from the summary provided above:

> Ozone is a "bad guy" in that breathing it is lethal at dosage levels of a few molecules per million air molecules. This is why ozone at the surface is referred to as a pollutant. Yet ozone high in the atmosphere screens out biologically harmful solar ultraviolet radiation, keeping it from reaching the surface. Such ultraviolet radiation is destructive of genetic cellular material in plants and animals, as well as human beings. Without the "ozone layer" high up in the atmosphere, life on the surface of the Earth would not be possible as we know it.
>
> [taken from the introduction to the ozone text]

Advance organizers:

- can be confusing if you mistake them for summaries
- can be too abstract
- can reference knowledge you don't have.

If you're having trouble:

- take your time reading and reflecting
- focus on what you do understand and try to come up with concrete examples that you're more familiar with
- give yourself a short break (meditate; rest; do something physical) before reading the full text.

Skimming

Once you've read any summaries provided, you're ready to lightly skim the text. **Skimming** is a widely misunderstood skill.

Skimming is **NOT**

- reading only the first and last words of a sentence

- reading very fast
- missing out the long words!

Skimming = actively searching text for critical information.

This requires you to pass your gaze lightly over the text, not reading word-by-word or even line-by-line, but taking a more global approach. So-called 'speed reading' courses do not teach you how to read faster, but (if they're any good) how to skim. Skimming is a very useful skill in certain circumstances:

- to find out whether a text is worth reading
- to quickly pick out the information of interest to you, from a simple text
- to prepare for studying a complex or difficult text.

Effective skimming requires you to pay attention to the cues in the text.

Texts provide several different types of cue:

- cues that signal content & organization: headings; summaries; 'signal' words
- cues that draw attention to particular ideas & details: highlighting (either with color or by bolding, underlining, or italicizing); questions; stated objectives
- cues that restate or elaborate a concept: examples; paraphrases; marginal comments; applications
- cues that relate text material to familiar information: comparisons; allusions to common experiences; quotes.

Part of being a skilled reader is being aware of the cues in the text.

Warning: Your better memory of signaled and highlighted information may be at the expense of other details. So if your goals are different from those of the author, and some details that are important to you are *not* signaled or highlighted, then you'll need to make a special effort to note them.

Headings

The best cues, if they're done well, are headings.

Good headings help your learning by:

- signaling what's important

- showing you how the text is organized
- helping you produce better summaries and outlines
- helping you remember the main points of the text.

A text with many good headings is also much easier to skim.

Look back at the ozone text. The original text has four, rather curt, section headings, which I removed (you can see them in the text below — they're the headings that are numbered). These do help break it up, but let's have a look at the text with more, and more informative, headings.

The Relationship Of Ozone And Ultraviolet Radiation: Why Is Ozone So Important?

In this section, we will explore what is ozone and what is ultraviolet radiation. We then will explore the relationship between ozone and ultraviolet radiation from the sun. It is here that ozone plays its essential role in shielding the surface from harmful ultraviolet radiation. By screening out genetically destructive ultraviolet radiation from the Sun, ozone protects life on the surface of Earth. It is for this reason that ozone acquires an enormous importance. It is why we study it so extensively.

2.1 Ozone and the Ozone Layer

Most ozone is in the stratosphere — the ozone layer

About 90% of the ozone in our atmosphere is contained in the stratosphere, the region from about 10 to 50-km (32,000 to 164,000 feet) above Earth's surface. Ten percent of the ozone is contained in the troposphere, the lowest part of our atmosphere where all of our weather takes place. Measurements taken from instruments on the ground, flown on balloons, and operating in space show that ozone concentrations are greatest between about 15 and 30 km.

How the ozone layer protects against harmful ultraviolet radiation

Although ozone concentrations are very small, typically only a few molecules O^3 per million molecules of air, these ozone molecules are vitally important to life because they absorb the biologically harmful ultraviolet radiation from the Sun. There are three different types of ultraviolet (UV) radiation, based on the wavelength of the radiation. These are referred to as UV-a, UV-b, and UV-c. UV-c (red) is entirely screened out by ozone around 35 km altitude, while most UV-a (blue) reaches the surface, but it is not as genetically damaging, so we don't worry about it too much. It is the UV-b (green) radiation that can cause sunburn and that can also cause genetic damage, resulting in things like skin cancer, if exposure to it is prolonged.

Ozone screens out most UV-b, but some reaches the surface. Were the ozone layer to decrease, more UV-b radiation would reach the surface, causing increased genetic damage to living things.

Ozone in the troposphere is a pollutant

Because most of the ozone in our atmosphere is contained in the stratosphere, we refer to this region as the stratospheric ozone layer. In contrast to beneficial stratospheric ozone, tropospheric ozone is a pollutant found in high concentrations in smog. Though it too absorbs UV radiation, breathing it in high levels is unhealthy, even toxic. The high reactivity of ozone results in damage to the living tissue of plants and animals. This damage by heavy tropospheric ozone pollution is often manifested as eye and lung irritation. Tropospheric ozone is mainly produced during the daytime in polluted regions such as urban areas. Significant government efforts are underway to regulate the gases and emissions that lead to this harmful pollution, and smog alerts are regular occurrences in polluted urban areas.

2.2 Solar Radiation

The sun produces radiation at various wavelengths

To appreciate the importance of stratospheric ozone, we need to understand something of the Sun's output and how it impacts living systems. The Sun produces radiation at many different wavelengths. These are part of what is known as the electromagnetic (EM) spectrum. EM radiation includes everything from radio waves (very long wavelengths) to X-rays and gamma rays (very tiny wavelengths). EM radiation is classified by wavelength, which is a measure of how energetic is the radiation. The energy of a tiny piece or "packet" of radiation (which we call a photon) is inversely proportional to its wavelength.

How we perceive different wavelengths

The human eye can detect wavelengths in the region of the spectrum from about 400 nm (nanometers or billionths of a meter) to about 700 nm. Not surprisingly, this is called the visible region of the spectrum. All the colors of light (red, orange, yellow, green, blue, and violet) fall inside a small wavelength band. Whereas radio waves have wavelengths on the order of meters, visible light waves have wavelengths on the order of billionths of a meter. Such a tiny unit is called a nanometer (1 nm= 10^{-9} m). At one end of the visible "color" spectrum is red light. Red light has a wavelength of about 630 nm. Near the opposite end of the color spectrum is blue light, and at the very opposite end is violet light. Blue light has a wavelength of about 430 nm. Violet light has a wavelength of about 410 nm. Therefore, blue light is more energetic than red light because of its shorter wavelength, but it is less energetic than violet light, which has an even shorter wavelength. Radiation with wavelengths shorter than those of violet light is called ultraviolet radiation.

Why ultraviolet wavelengths are dangerous

The Sun produces radiation that is mainly in the visible part of the electromagnetic spectrum. However, the Sun also generates radiation in ultraviolet (UV) part of the spectrum. UV wavelengths range from 1 to 400 nm. We are concerned about ultraviolet radiation because these rays are energetic enough to break the bonds of DNA molecules (the molecular carriers of our genetic coding), and thereby damage cells. While most plants and animals are able to either repair or destroy damaged cells, on occasion, these damaged DNA molecules are not repaired, and can replicate, leading to dangerous forms of skin cancer (basal, squamous, and melanoma).

2.3 Solar Fluxes

Solar energy is measured by 'solar flux'

Solar flux refers to the amount of solar energy in watts falling perpendicularly on a surface one square centimeter, and the units are watts per cm^2 per nm. Because of the strong absorption of UV radiation by ozone in the stratosphere, the intensity decreases at lower altitudes in the atmosphere. In addition, while the energy of an individual photon is greater if it has a shorter wavelength, there are fewer photons at the shorter wavelengths, so the Sun's total energy output is less at the shorter wavelengths. Because of ozone, it is virtually impossible for solar ultraviolet to penetrate to Earth's surface. For radiation with a wavelength of 290 nm, the intensity at Earth's surface is 350 million times weaker than at the top of the atmosphere. If our eyes detected light at less than 290 nm instead of in the visible range, the world would be very dark because of the ozone absorption!

2.4 UV Radiation and the Screening Action by Ozone

Damage potential of radiation is measured by the action spectrum

To appreciate how important this ultraviolet radiation screening is, we can consider a characteristic of radiation damage called an action spectrum. An action spectrum gives us a measure of the relative effectiveness of radiation in generating a certain biological response over a range of wavelengths. This response might be erythema (sunburn), changes in plant growth, or changes in molecular DNA. Fortunately, where DNA is easily damaged (where there is a high probability), ozone strongly absorbs UV. At the longer wavelengths where ozone absorbs weakly, DNA damage is less likely. If there was a 10% decrease in ozone, the amount of DNA damaging UV would increase by about 22%. Considering that DNA damage can lead to maladies like skin cancer, it is clear that this absorption of the Sun's ultraviolet radiation by ozone is critical for our well-being.

Different types of UV radiation

While most of the ultraviolet radiation is absorbed by ozone, some does make it to Earth's surface. Typically, we classify ultraviolet radiation into three parts, UV-a (320-400 nm), UV-b (280-320 nm), and UV-c (200-280 nm). Sunscreens have been developed by commercial manufacturers to protect human skin from UV radiation. The labels of these sunscreens usually note that they screen both UV-a and UV-b. Why not also screen for UV-c radiation? When UV-c encounters ozone in the mid-stratosphere, it is quickly absorbed so that none reaches Earth's surface. UV-b is partially absorbed and UV-a is barely absorbed by ozone. Ozone is so effective at absorbing the extremely harmful UV-c that sunscreen manufacturers don't need to worry about UV-c. Manufacturers only need to eliminate skin absorption of damaging UV-b and less damaging UV-a radiation.

How time and season affect how much UV radiation is absorbed by ozone

The screening of ultraviolet radiation by ozone depends on other factors, such as time of day and season. The angle of the Sun in the sky has a large effect on the UV radiation. When the Sun is directly overhead, the UV radiation comes straight down through our atmosphere and is only absorbed by overhead ozone. When the Sun is just slightly above the horizon at dawn and dusk, the UV radiation must pass through the atmosphere at an angle. Because the UV passes through a longer distance in the atmosphere, it encounters more ozone molecules and there is greater absorption and, consequently, less UV radiation striking the surface.

No question which text is most helpful in selecting what's important in the text!

Why do headings help memory?

- ➤ If you have knowledge of the topic, headings are useful for priming (bringing that knowledge to mind; preparing you for the text so you can process it better).

- ➤ Headings are useful for helping you search for specific information.

- ➤ Headings help you scan the text quickly, to get the 'big picture'.

- ➤ Good headings facilitate transitions.

One reason that stories are so much easier to remember is that they follow a chain of causal events: this happens, so this happens.

How difficult an **expository text** is to understand and remember depends a lot on the extent to which the information flows in a logical and predictable order. Clearly written expository text handles the transitions from one topic to another well.

The difficulty of a text, then, is directly related to:

- how clear its structure is to the reader, and
- how well it manages the transitions between topics.

How to use headings to help deal with difficult texts

Use headings

- to help you understand how the text is organized
- as a framework for your notes
- as signs to what is most important.

Write your own headings in when a text:

- has too many topics
- has too much detail
- is poorly organized.

Add headings if:

- there's too much detail under the existing heading
- all transitions aren't given a heading.

Re-write existing headings when they:

- don't clearly express the main idea of the section
- don't express what's most important in the section *for you*.

Identifying text structure

Unlike narrative texts (stories), expository texts can follow many different patterns. Good readers are sensitive to these **text structures**.

Text structure:

> - cues you to what is important
> - helps you ask the right questions
> - tells you how the ideas in the text are related
> - provides a structure that helps you form a useful mental model
> - cues you to the best format for your notes.

Common types of structure used in expository texts include:

> - **Description/Generalization**
> - **Collection/Enumeration**
> - **Cause & effect**
> - **Sequence**
> - **Classification**
> - **Comparison / contrast**
> - **Problem**
> - **Refutational**

Signal words provide cues to help you identify the specific text structure.

Description/Generalization

extends or clarifies main ideas through explanations, examples, or information about attributes.

In this type of structure, a paragraph always has a main idea. Other sentences in the paragraph either clarify the main idea by giving examples or illustrations, or extend the main idea by explaining it in more detail. Here's an example:

> Comprehension is defined as ... Thus, readers derive meaning from text when ... The data suggest that that text comprehension is enhanced when ...

Elements: main idea; key words; examples

Common signal words: *this is a story about, this report will describe, in particular, for instance, for example, is defined, to illustrate, characteristics*

Collection/Enumeration

lists facts or elements.

Enumeration passages may be a bulleted or numbered list, or a list of items in paragraph form. For example:

> Reading is a complex skill that involves several levels of decoding and interpretation:
>
> 1. the physical features of the letters
> 2. the words
> 3. the meaningful chunks of phrases
> 4. ...

Elements: topic; subtopics; details

Common signal words: *in addition to, and ... and ... and, a number of, many, one... two ...three, and so forth*

Cause & Effect

lists causes or events and the resulting effects or consequences. For example:

> Because of the strong absorption of UV radiation by ozone in the stratosphere, the intensity decreases at lower altitudes in the atmosphere. In addition, ..., so the Sun's total energy output is less at the shorter wavelengths. Because of ozone, it is virtually impossible for solar ultraviolet to penetrate to Earth's surface

Elements: causes; effects

Common signal words: *reasons for, reasons why, if ... then, as a result of, therefore, because, since, leads to, since, result, outcome, impact, influenced by, brought about by, consequently*

Sequence

describes a connecting series of events or steps, possibly causally related.

For example:

> When the right atrium is filled with blood ... which then ... is then carried by the pulmonary veins ... From whence, ... , and then leaves the heart ... From there, it ...

Elements: topic; steps; details

Common signal words: *first ... second ..., finally, therefore, consequently, the next, previously, before, after, if ... then, leads to, causes, because*

Getting an Understanding of the Text

Classification

groups items into classes or categories. For example:

> The brain contains two types of nerve cell: *neurons* and *glia*. ... Neurons, like glia, are a broad class of cells. There are several different types of neuron, but they all have certain attributes in common.

Elements: topic; categories; details

Common signal words: *belongs to, types, classes, categories, group*

Comparison / contrast

examines the relationships between two or more things.

In comparison, both similarities and differences are studied. In contrast, only the differences are noted.

For example:

> The researchers then measured brain activity in two separate groups ... Functional magnetic resonance imaging (fMRI) scans revealed that those with one copy of "met" showed ..., in contrast to lack of activation among those with two copies of "val".

Elements: topic; items; differences &/or similarities

Common signal words:

Comparison: *likewise, similarly, same, in comparison, alike, in the same way, just as*

Contrast: *in contrast, but, however, nevertheless, on the other hand, on the contrary, different*

Problem

discusses a problem and its solution, or a question and its answer.

This type of text can resemble a cause-&-effect structure. Typically, a text with this structure will state a problem and its causes and effects, then suggest one or more solutions; sometimes solutions might be suggested first, followed by problem, causes, effects. For example:

> The problem of stress today ... Many factors contribute to stress ... The consequences of prolonged stress include ... How can we reduce the problem of stress ...

Elements: problem; causes; effects; solutions

Common signal words: *problem, dilemma, trouble, difficulty, puzzle, if...then, because, so that, question/answer, solution*

Refutational

addresses and refutes a common misconception.

Typically, a text with this structure will contain a statement describing a common misconception and a direct statement contradicting this belief.

> The whole idea of right brain vs left brain ..., but ... the myth that developed ... It is true that ... But ...

Elements: misconception; reason; facts; evidence

Common signal words: *some people believe, some people think, a common myth, a common misconception, this is not true*

To identify the type of text structure, look not only for signal words and elements, but also at the overall theme of the text. Note that the signal words provided are not a complete set! Note, too, that signal words are pointers only. You will have noticed that some words appear in more than one list, and some words are so common that their occasional presence can hardly be a definite sign of a particular structure. Signal words point you in a direction, but you have to consider more than that to determine a text's structure.

You will also have noticed that the types of text structure seem to overlap. A text will often have elements of more than one structure; which one you choose for your organization will depend in part on what the most important elements are to you. To know this, you will need to have identified your goal.

In the section on note-taking, we will look at how text structure guides your notes.

Illustrations

Illustrations can dramatically improve your understanding and **recall** of the material.

Most students give illustrations little of the attention they need to be effective.

"Illustrations" include:

Pictures

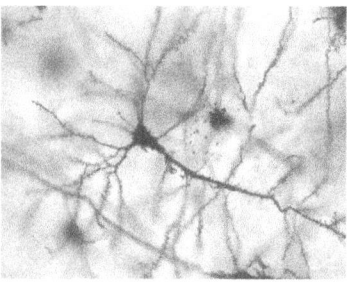

Graphs

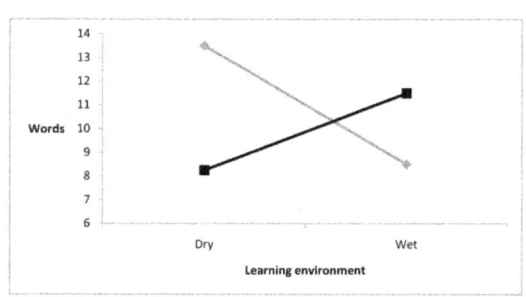

Charts

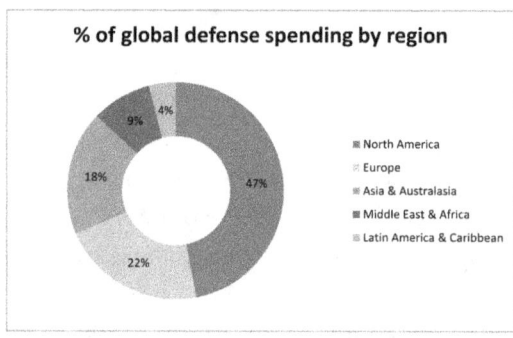

Flowcharts

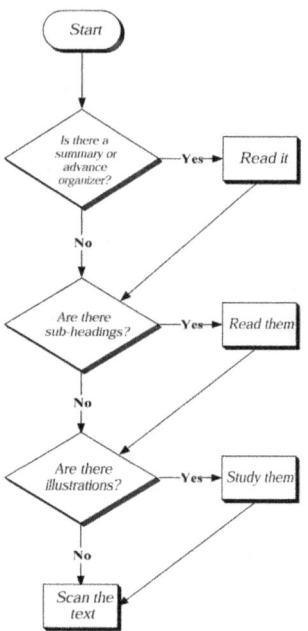

Timelines

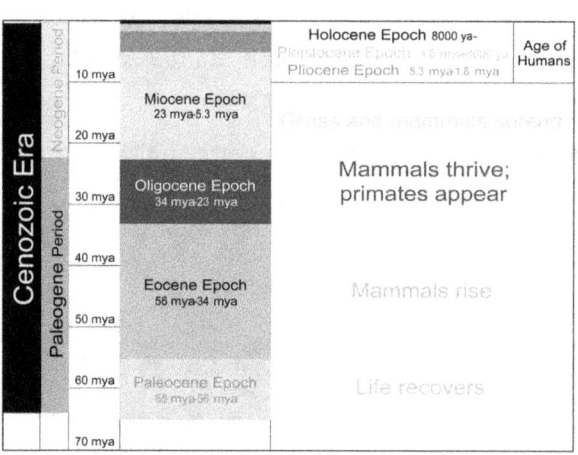

Diagrams

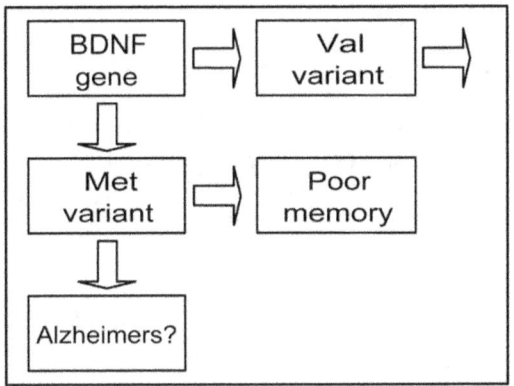

Maps

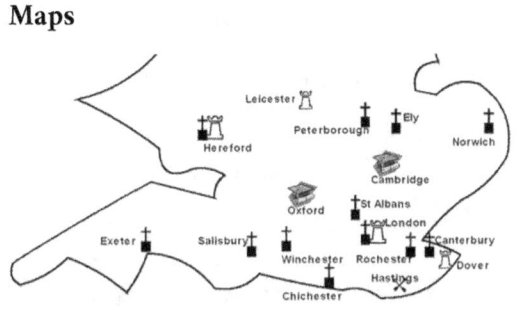

Tables

Eon	Era	Event Marker
Phanerozoic Eon 542 my - Present	Cenozoic Era 65 my - present	K-T Extinction 65 my
	Mesozoic Era 250 my - 65 my	Permian Extinction 250 my
	Paleozoic Era 542 my - 250 my	Cambrian Explosion 542 my
Precambrian Eon 4565 my - 542 my		

Multimedia summaries

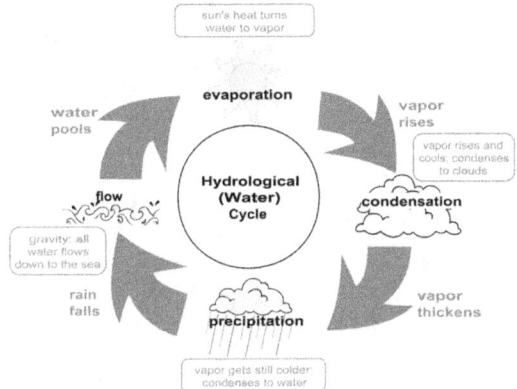

Why are illustrations valuable?

Illustrations can help you understand and remember text by:

➢ providing a visual image of the information to be remembered

➢ providing an organizing scheme

➢ motivating you to spend more time and effort in studying the material

➢ providing more depth or **elaboration** to the material

➢ providing a context for the material

➢ activating prior knowledge

➢ presenting information in a more scannable form

➢ displaying relationships and important information more obviously.

But all illustrations aren't equally valuable. To benefit most from illustrations, you need to know which ones are worth attending to.

Categorizing illustrations accordingly to their mnemonic value

There are 5 functions an illustration may serve; here they are ordered according to the degree to which they may help memory, with the least helpful first:

1. **decoration**

2. **representation** (illustrative of the accompanying text, as in a child's illustration book)

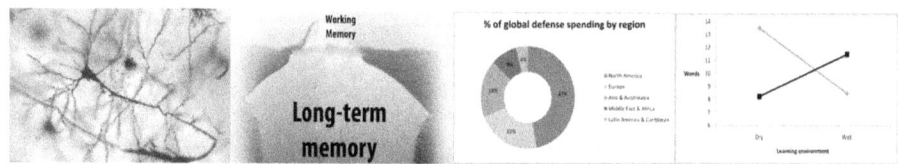

3. **organization** (to provide a framework for the text; e.g., illustrated maps, how-to-do-it diagrams)

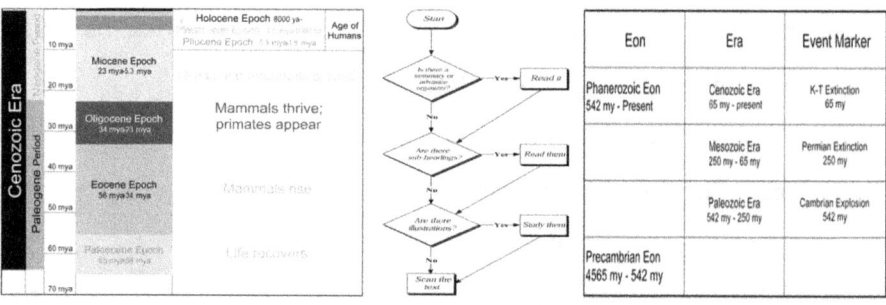

4. **interpretation** (to clarify hard-to-understand passages and abstract concepts within passages; common in science and social science)

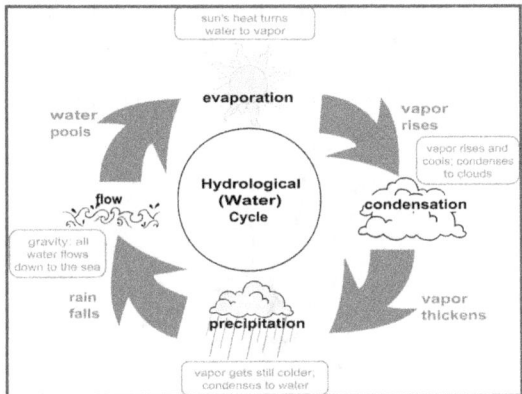

5. **transformation** (those that elaborate particular details by providing a mnemonic)

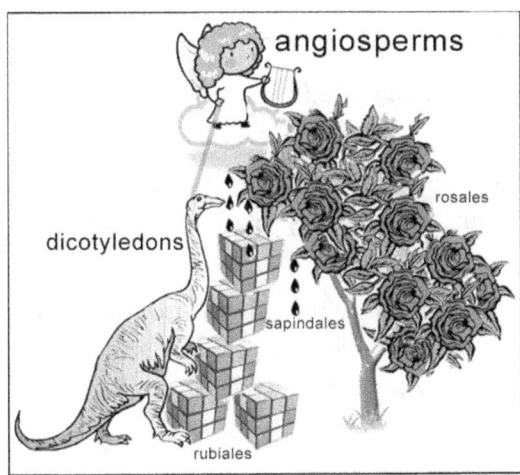

The most useful types of illustration are the last three:

> An organizational illustration improves memory by helping you organize the material.

> An interpretational illustration improves memory by helping you understand the text.

> A transformational illustration improves memory by giving you a visual image that helps you remember specific details, usually names.

Decorative illustrations are of little mnemonic value.

Representational illustrations draw more attention to the text, and thus can help the text be better remembered, but they do so at the expense of text that is not illustrated. Representational illustrations are most useful when:

> ➢ they provide a needed visual discrimination (such as identifying parts of a heart):

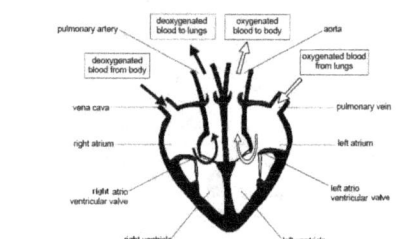

Diagram of heart, by J. R. Lawson, Otago Polytechnic, Dunedin, New Zealand, 2008.

> ➢ they illustrate spatial-structural relationships (as this heart diagram also does)
> ➢ they illustrate main ideas rather than details
> ➢ the text can't be understood without reference to them.

In general, 'pictures' usually serve a decorative or representational function, while 'graphics' (e.g., graphs, maps, diagrams, flowcharts, timelines, tables and charts) usually serve an organizational, interpretative, or transformational function.

However, the difference between representation, organization, and interpretation, can be quite fuzzy, and which of these functions is primary depends in part on the complexity of the text and your own knowledge and reading skill:

> ➢ Illustrations are redundant if you can easily understand and remember the text.
> ➢ Most illustrations are equally useless if the text is too difficult for you.
> ➢ Illustrations are most useful when you need a little help, but only a little, to understand the text.

The level of detail in the illustration also affects how useful it is. The most useful amount of detail is affected by:

> ➢ your level of relevant knowledge

- your ability to process detailed illustrations
- the amount of time you have to study it (if time is short, line drawings are usually better; if you have more time, you can handle more complex illustrations)
- the function the illustration serves (e.g., a representational illustration should focus on the main ideas, while a transformational illustration should focus on a very few specific details — maybe only one)
- the action the illustration takes (this list is ordered according to how common these actions are; the first is by far the most common, and the last quite rare):
  - ★ **reinforces**: simply repeats the information described in the text
  - ★ **summarizes** (this can be difficult to distinguish from the reinforce function)
  - ★ **embellishes**: provides additional information to what is described in the text
  - ★ **elaborates**: repeats some of the information in the text and add to it
  - ★ **compares**: enables the reader to compare the information to information contained in an earlier graphic.
- the presence and quality of a caption: a good caption draws attention to specific details in the illustration and helps you interpret it.

Compare these two graphs, the first without a caption and the second with one:

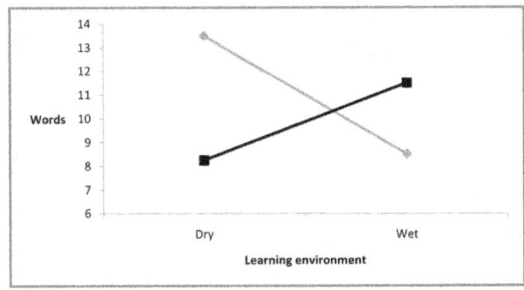
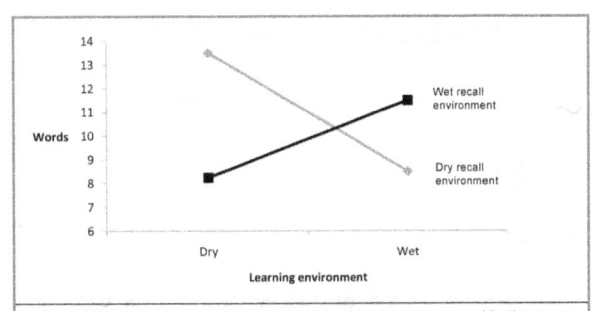

A list of 40 unrelated words were remembered better when tested in the same environment (on land or 20 feet underwater) as they had been learned in (Gordon & Baddeley 1975)

Getting an Understanding of the Text

Graphics

As with text structure, specific types of graphic are most appropriate for different types of information. Understanding this will also help you judge

- whether a particular illustration is worth spending time on,
- how much effort it's worth, and
- where that effort should be directed.

Multimedia summaries

Multimedia summaries combine pictures and words. The close integration of text and image is critical! In the water cycle illustration below, notice how:

- the words are tightly integrated with the images
- each image has a one-word label plus a brief description
- each arrow has its own explanatory label
- the pictures would lose much of their value without the words
- the words would lose a significant amount of their memorability without the images.

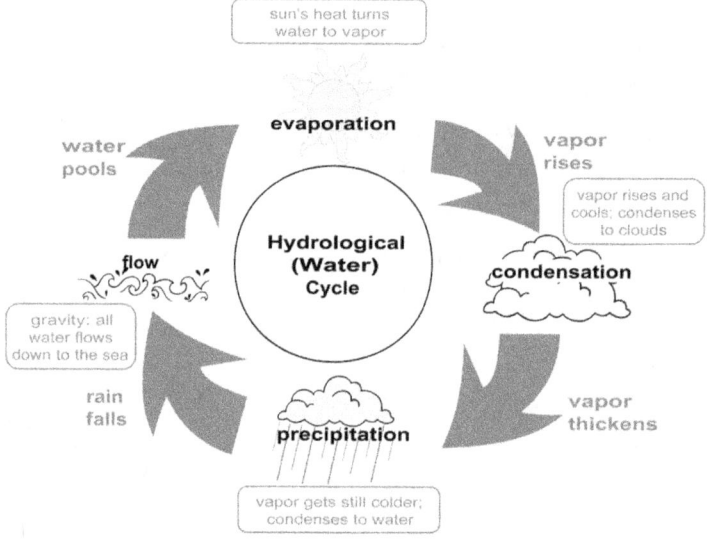

Multimedia summaries are the most help when:

- the text contains a cause-and-effect chain rather than a list of facts

- the illustration contains a series of frames showing the steps, together with integrated labels, rather than a one-frame static illustration
- text is minimal
- your goal is understanding, not rote regurgitation.

The best multimedia summaries are:

- **concise**: there's a small number of simple illustrations and only a small number of words
- **coherent**: illustrations are presented in a cause-and-effect sequence
- **coordinated**: text is presented together with the relevant illustration.

Why? Because all this reduces the load on working memory.

Multimedia summaries may be of particular benefit:

- to those with a low working memory capacity
- when your working memory is taken up with other things (such as anxiety about something, or a simultaneous activity)
- when the text as a whole is demanding.

Diagrams

Diagrams do two things well:

- isolate important points
- display directional flow (most often seen in a cause-effect relationship).

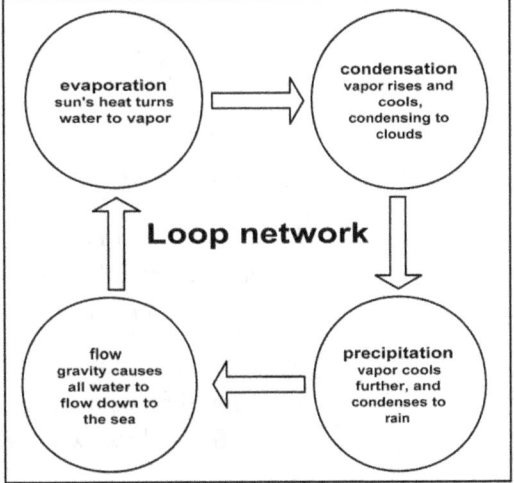

To a large extent, diagrams help in the same way that multimedia summaries do. The main difference between the two categories is that multimedia summaries include pictorial elements. Compare the previous multimedia summary with the diagram for the same information.

A diagram is generally simpler than a multimedia summary, and thus may be quicker to process. However, this is not necessarily an advantage! A 'busier' illustration, such as the multimedia summary above, is both more memorable and more comprehensible, even as it encourages you to spend a little more time on it.

Maps

In general, humans have very good spatial memory, which is why remembering how items are laid out often helps us recall the items. Maps can therefore help memory considerably.

By helping you direct your attention to what matters, maps are better studied *before* the text.

To use a map effectively, you should focus most on the relationships *between* features, rather than the specific features themselves.

Again, it's all about cognitive load:

> ➤ A useful map has only a limited number of features (12-16).

> ➤ Icons (simple representative pictures) are usually better than arbitrary geometric symbols.

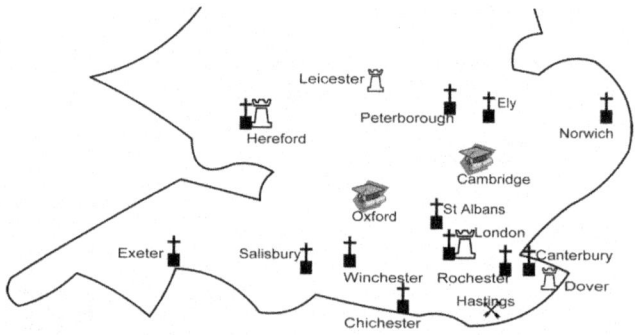

Timelines

Timelines may, in a sense, be thought of as maps of time. This is not as strange as you might think. We perceive time in spatial terms, in terms of distance and direction. Timelines transform dates into a spatial array that helps us remember.

As with maps, then, timelines are usefully studied *before* reading the text, and you should focus most of your attention on the temporal relationships between events (how close; whether before or after; whether contained within).

Many timelines will be one simple line; the one below is more complex. Here, different sections contain other sections, and these contain still more. Note too, that because of the vast 'distance' covered, the most recent epochs are too (relatively) small to contain their names and dates, and have had to be written beside.

Cenozoic Era	Neogene Period	10 mya		Holocene Epoch 8000 ya–	Age of Humans
				Pleistocene Epoch 1.8 mya–8000 ya	
				Pliocene Epoch 5.3 mya–1.8 mya	
		20 mya	Miocene Epoch 23 mya–5.3 mya	Grass and mammals spread	
	Paleogene Period	30 mya	Oligocene Epoch 34 mya–23 mya	Mammals thrive; primates appear	
		40 mya			
		50 mya	Eocene Epoch 56 mya–34 mya	Mammals rise	
		60 mya	Paleocene Epoch 65 mya–56 mya	Life recovers	
		70 mya			

With a complex timeline like this, you might need to approach it in sections — studying the first three columns initially, before moving on to the fourth (including the insert that appears in the fifth column but is not part of it), and finally the fifth. In your initial foray, you would note that the timeline is covering the Cenozoic Era, and that this includes two Periods: the Paleogene and the Neogene; that the Cenozoic Era covers a period of some 65 million years, and the split between the two component Periods occurs at around 23 million years. Studying the fourth column, you would note how the epochs relate to the Periods in the second column, and how they relate to each other (their order). With the fifth column, you would briefly note the order of events, then attend to how they relate to the epochs in the fourth column, to the Periods in the second column, and the Era in the first (i.e., that the Cenozoic Era begins with life recovering and is marked by the rise of mammals, and ultimately humans).

Tables

Tables are usually easier and quicker to process than other types of graphics. They provide a quick reference for numerical data or simple hierarchical facts, allowing you to see certain relationships between the items. Compare the timeline with the table below.

Era	Period	Epoch	Event Marker
Cenozoic Era 65 my - Present	Neogene Period 23 my - present	Holocene Epoch 8000 ya - present	Human history
		Pleistocene Epoch 1.8 my - 8000 ya	Great Ice Age
		Pliocene Epoch 5.3 my - 1.8 my	Humans arise
		Miocene Epoch 23 my - 5.3 my	Grass & mammals spread
	Paleogene Period 65 my - 23 my	Oligocene Epoch 34 my - 23 my	Primates appear
		Eocene Epoch 56 my - 34 my	Mammals rise
		Paleocene Epoch 65 my - 56 my	Life recovers
			K-T Extinction 65 my

This is an easier display to process. You can see at a glance the names and dates of the various time periods, and you can easily see which epochs belong to which periods, and that they all belong to the Cenozoic Era.

The information is a little harder to process in the timeline, but the timeline has two advantages:

> The more visual display makes it easier to remember.

> It gives a sense of the relative sizes of the time periods (information which can only be derived from the table by mathematical calculation).

Graphs & Charts

Graphs vary considerably in their complexity, but don't be fooled by apparent simplicity! Even the simplest (*especially* the simplest) may be hard to understand, or even misleading.

When confronted by a graph, ask yourself: What story is the writer trying to convey?

Graphs are selective tools, that are there to portray and emphasize specific facts or relationships.

For example, look at this pie chart:

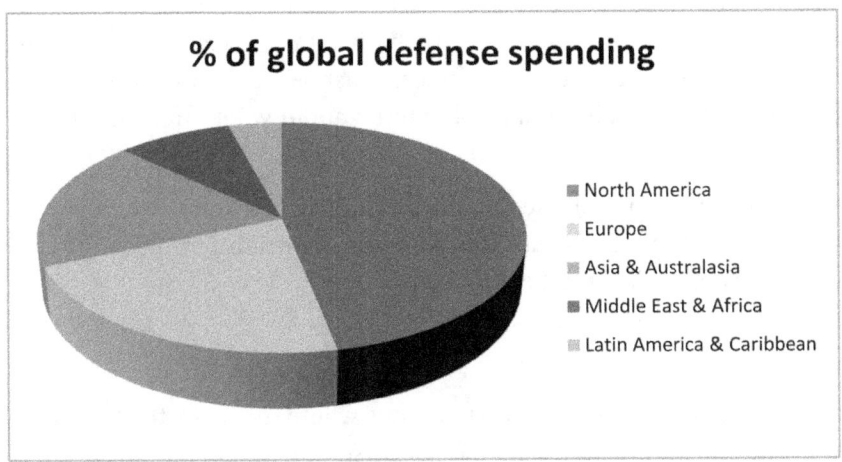

Pie charts are notoriously poor at accurately conveying relative size. This one, for example, seems to show that Europe spends more than half as much as North America on defense, when it's actually *less* than half (22% vs 47%).

The particular partitioning also is a way of trying to de-emphasize the contribution of the U.S. See how the second graph allows more and better nation comparisons.

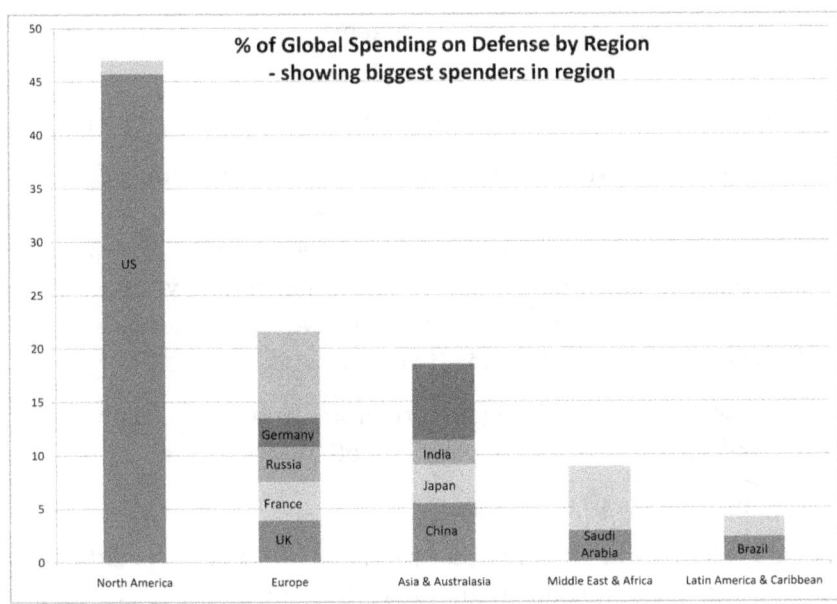

Getting an Understanding of the Text 67

Don't think that all this is simply because of the political aspect of this information! Even scientific graphs are there to tell a particular story, make a specific point. There is nothing wrong with this; it is inescapable: graphs are limited in what they can portray, so choices must be made. But to 'read' the graph, to understand it fully, you need to think about what information the writer is trying to convey.

This not only helps you evaluate the information, it also helps you in your primary task: to determine what information is important. Information that is 'pushed' in a graph is likely to be important.

Checklist

To assess the value of an illustration (and thus whether it's worth attending to, and how much effort it rates), ask yourself these questions:

1. What kind of illustration is it? (picture; map; graph; diagram; table; etc)

2. What function does it serve? (decorative; representational; organizational; interpretative; transformational)

    if representational: does it add important or clarifying detail to the text? does it portray structural or discriminatory detail? does it point to what is important in the text?

    if organizational: does it help *you* organize the information in a way that you find useful?

    if interpretative: does it help *you* understand the material better?

    if transformational: do you already have this information well-learned? do you regard it as unnecessary, either because you believe the information is not worth learning or because you don't think it will help you remember it? (If so, think again! despite your beliefs, the odds are that this information is both worth memorizing, and that using it will help you more than you think).

3. Which action does it take? Does it reinforce, summarize, embellish, or elaborate the text, or compare to another illustration?

4. How difficult is it to understand?

Like text, graphics vary in difficulty, and your ability to assess the difficulty of a graphic helps you know how much time and effort will be needed to understand it.

How to assess the difficulty of a graphic illustration:

- **degree of abstraction**: the more abstract, the more difficult

  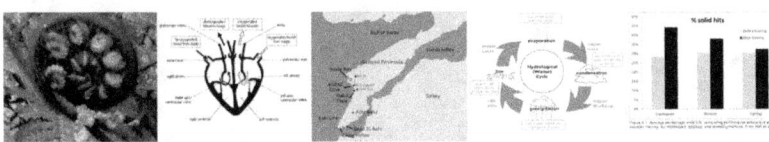

  Illustrations vary in their degree of abstraction, with photographs the most concrete and realistic, and graphs the most abstract.

- **presence and accuracy of caption**: the clearer the caption, the easier it is to understand the graphic
- **contextual detail**: contextual details help you understand — beware of illustrations that are over-simplified; many high school textbooks, for example, "clean up" graphs, removing details such as the units used, supposedly to make them easier to read, but actually making them more difficult to understand
- **detailed explanation and description**: even experts in a topic sometimes have trouble understanding textbook illustrations, because the illustrations lack the details that make sense of them
- **distance from relevant text**: if illustrations are placed on different pages than their accompanying text, it becomes much more difficult to relate the two together
- **grouping**: illustrations are often best understood when considered together with other, related, illustrations; however textbooks often separate or even omit these, making it harder to understand them.

How to use illustrations effectively

- Use illustrations as you would text: as a potential source of relevant information.
- Give illustrations appropriate time and attention (don't ignore them, but don't give them more time and effort than they're worth).
- To direct your efforts, try doing specific activities, such as:
  - ★ trying to answer questions that direct attention to particular information
  - ★ labeling features on a map

★ labeling parts on a illustration

★ "tracing" a photo for five minutes.

Text

Traditional advice when writing an essay is that you begin each paragraph with the 'topic sentence' (a sentence explaining the main idea that will be discussed in the paragraph) and end with a summary or concluding sentence, summing up the material you've covered.

This basic format is the reason why the first and last sentences of paragraphs in expository text are more likely to contain the important information. This is *not* to say that important information won't be found elsewhere! But skimming is not about getting all the information you need; it's about getting a picture of the material covered. Paying more attention to the first or last sentences (or both, if paragraphs are long) will give you a better feel for the material covered, and will also give you a better sense of the transitions (how well the text flows from one idea to another).

Here, for example, are the first sentences of each paragraph in the ozone text:

> In this section, we will explore what is ozone and what is ultraviolet radiation.
>
> About 90% of the ozone in our atmosphere is contained in the stratosphere, the region from about 10 to 50-km (32,000 to 164,000 feet) above Earth's surface.
>
> Although ozone concentrations are very small, typically only a few molecules O^3 per million molecules of air, these ozone molecules are vitally important to life because they absorb the biologically harmful ultraviolet radiation from the Sun.
>
> Because most of the ozone in our atmosphere is contained in the stratosphere, we refer to this region as the stratospheric ozone layer. .
>
> To appreciate the importance of stratospheric ozone, we need to understand something of the Sun's output and how it impacts living systems.
>
> The human eye can detect wavelengths in the region of the spectrum from about 400 nm (nanometers or billionths of a meter) to about 700 nm.
>
> The Sun produces radiation that is mainly in the visible part of the electromagnetic spectrum.

Solar flux refers to the amount of solar energy in watts falling perpendicularly on a surface one square centimeter, and the units are watts per cm^2 per nm.

To appreciate how important this ultraviolet radiation screening is, we can consider a characteristic of radiation damage called an action spectrum.

While most of the ultraviolet radiation is absorbed by ozone, some does make it to Earth's surface.

The screening of ultraviolet radiation by ozone depends on other factors, such as time of day and season.

Compare this with the set of final sentences:

It is for this reason that ozone acquires an enormous importance. It is why we study it so extensively.

Measurements taken from instruments on the ground, flown on balloons, and operating in space show that ozone concentrations are greatest between about 15 and 30 km.

Were the ozone layer to decrease, more UV-b radiation would reach the surface, causing increased genetic damage to living things.

Significant government efforts are underway to regulate the gases and emissions that lead to this harmful pollution, and smog alerts are regular occurrences in polluted urban areas.

The energy of a tiny piece or "packet" of radiation (which we call a photon) is inversely proportional to its wavelength.

Radiation with wavelengths shorter than those of violet light is called ultraviolet radiation.

While most plants and animals are able to either repair or destroy damaged cells, on occasion, these damaged DNA molecules are not repaired, and can replicate, leading to dangerous forms of skin cancer (basal, squamous, and melanoma).

If our eyes detected light at less than 290 nm instead of in the visible range, the world would be very dark because of the ozone absorption!

Considering that DNA damage can lead to maladies like skin cancer, it is clear that this absorption of the Sun's ultraviolet radiation by ozone is critical for our well-being.

Manufacturers only need to eliminate skin absorption of damaging UV-b and less damaging UV-a radiation.

> Because the UV passes through a longer distance in the atmosphere, it encounters more ozone molecules and there is greater absorption and, consequently, less UV radiation striking the surface.

As you can see, the set of first sentences gives a better idea for the flow of ideas. However, some important ideas only appear in the last sentences.

Here is the complete set of both first and last sentences:

> In this section, we will explore what is ozone and what is ultraviolet radiation.
>
> It is for this reason that ozone acquires an enormous importance. It is why we study it so extensively.
>
> About 90% of the ozone in our atmosphere is contained in the stratosphere, the region from about 10 to 50-km (32,000 to 164,000 feet) above Earth's surface.
>
> Measurements taken from instruments on the ground, flown on balloons, and operating in space show that ozone concentrations are greatest between about 15 and 30 km.
>
> Although ozone concentrations are very small, typically only a few molecules O^3 per million molecules of air, these ozone molecules are vitally important to life because they absorb the biologically harmful ultraviolet radiation from the Sun.
>
> Were the ozone layer to decrease, more UV-b radiation would reach the surface, causing increased genetic damage to living things.
>
> Because most of the ozone in our atmosphere is contained in the stratosphere, we refer to this region as the stratospheric ozone layer.
>
> Significant government efforts are underway to regulate the gases and emissions that lead to this harmful pollution, and smog alerts are regular occurrences in polluted urban areas.
>
> To appreciate the importance of stratospheric ozone, we need to understand something of the Sun's output and how it impacts living systems.
>
> The energy of a tiny piece or "packet" of radiation (which we call a photon) is inversely proportional to its wavelength.
>
> The human eye can detect wavelengths in the region of the spectrum from about 400 nm (nanometers or billionths of a meter) to about 700 nm.
>
> Radiation with wavelengths shorter than those of violet light is called ultraviolet radiation.

The Sun produces radiation that is mainly in the visible part of the electromagnetic spectrum.

While most plants and animals are able to either repair or destroy damaged cells, on occasion, these damaged DNA molecules are not repaired, and can replicate, leading to dangerous forms of skin cancer (basal, squamous, and melanoma).

Solar flux refers to the amount of solar energy in watts falling perpendicularly on a surface one square centimeter, and the units are watts per cm^2 per nm.

If our eyes detected light at less than 290 nm instead of in the visible range, the world would be very dark because of the ozone absorption!

To appreciate how important this ultraviolet radiation screening is, we can consider a characteristic of radiation damage called an action spectrum.

Considering that DNA damage can lead to maladies like skin cancer, it is clear that this absorption of the Sun's ultraviolet radiation by ozone is critical for our well-being.

While most of the ultraviolet radiation is absorbed by ozone, some does make it to Earth's surface.

Manufacturers only need to eliminate skin absorption of damaging UV-b and less damaging UV-a radiation.

The screening of ultraviolet radiation by ozone depends on other factors, such as time of day and season.

Because the UV passes through a longer distance in the atmosphere, it encounters more ozone molecules and there is greater absorption and, consequently, less UV radiation striking the surface.

Reading both first and last sentences gives you a much fuller sense of the content, but of course, it doubles the time and effort, and includes quite a lot of redundant information. Given that, I recommend only reading both:

➢ if you're not planning to read the text properly, or

➢ you have little background knowledge of the subject, or

➢ the text is difficult.

Main Points

Before reading the text, get a big picture view:

- Read any summary or advance organizer first
    - use it as a priming tool (to trigger relevant knowledge you already have)
    - use it to help you frame your initial goal (what you hope to learn from the text)
    - take from it an idea of what information the writer believes is most important
    - try and condense the most important idea(s) from the summary into one sentence, in your own words.
- If you're having trouble doing this, focus on what you do understand, and reflect on that. Then take a break before coming back to it. Go through the summary again, but if you're still finding it difficult, move on to the next step. Take your difficulty as a sign that you need to go slowly.
- Lightly skim the text, looking out for the cues in the text. Don't try to do too much at a time. If the text is difficult, take several passes, focusing on different types of cue.
    - Scanning the headings gives you an idea of how the text is organized.
    - Signal words suggest the type of text structure.
    - Words that are highlighted, the first and last sentences in a paragraph, and explicit questions and objectives, all signal information that's considered important.
    - Examples, comparisons, and illustrations, all elaborate the concepts in the text (i.e., add to ideas in ways that can help improve and deepen your understanding).

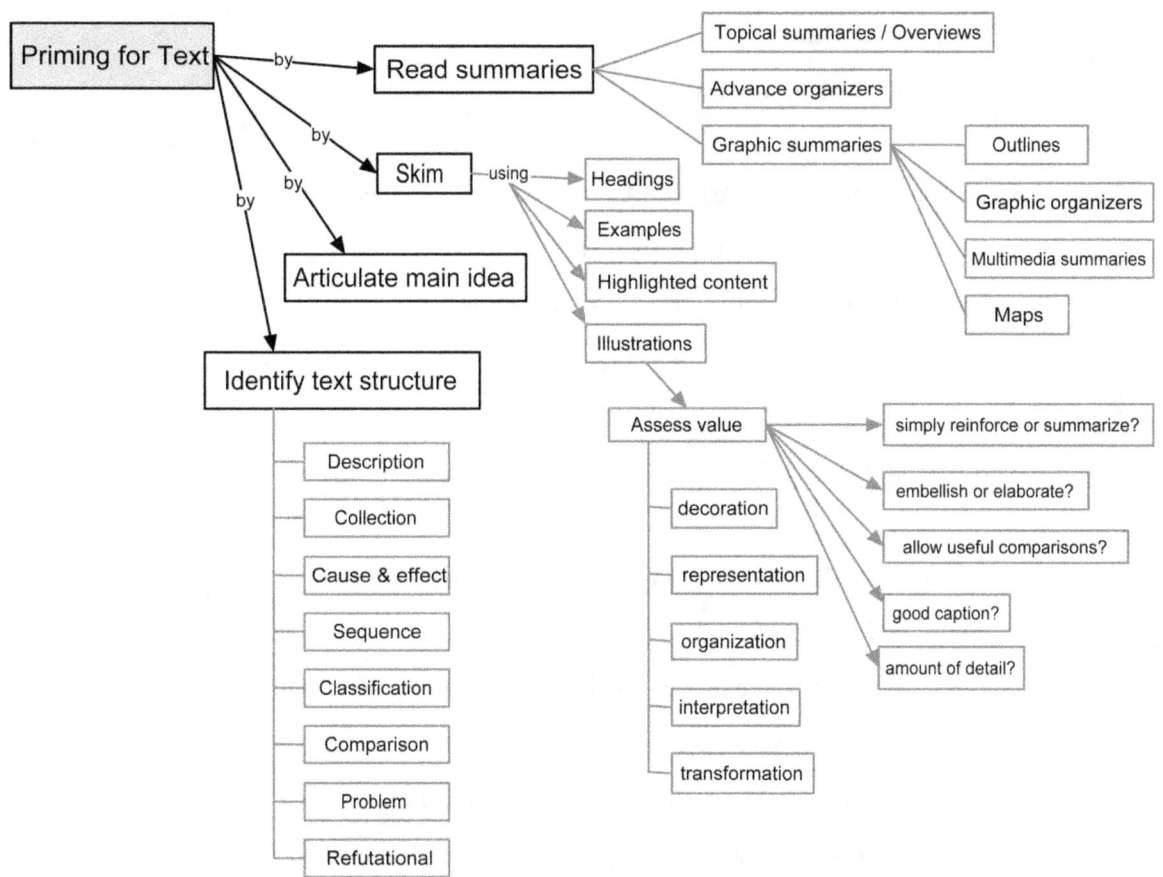

Getting an Understanding of the Text

Review Questions

1. The "Main Points" found at the end of this chapter are an example of:

    a. Overview

    b. Advance organizer

    c. Outline

    d. Topical summary

2. Topical summaries and overviews help you learn by:

    a. Preparing you for the text

    b. Giving an idea of what's important

    c. Giving you ready-made notes

    d. Helping you remember what's important

    e. Making the text faster to read and understand

3. Advance organizers can be recognized by the following characteristics:

    a. Placed before the text

    b. Written at a higher level of abstraction

    c. List the main points

    d. Focus on the most important information

    e. Include details or perspective that are not in the text

4. Skimming is a form of reading that:

    a. Skips all the long or difficult words

    b. Omits all but the first and last line of each paragraph

    c. Is focused on actively seeking important information

    d. Is useful for assessing whether a text is worth reading

5. Select the features in the text that are useful when skimming:

   a. Organizational illustrations

   b. Signal words

   c. Decorative illustrations

   d. Headings

   e. Overviews

   f. Topic sentences

6. Headings are useful for:

   a. Helping you work out what's important

   b. Skimming

   c. Showing how the text is organized

   d. Helping you remember the main points

7. Why is it helpful to assess the value of an illustration?

   a. Knowing the kind of illustration helps you know how much time it's worth

   b. Knowing its function helps you ask the right questions about its value

   c. Knowing what action it takes tells you how much it will help you remember the information in the text

   d. Knowing its action and function helps you direct your efforts appropriately

8. Identifying the structure of an expository text helps you:

   a. Organize your thoughts

   b. Work out what's important

   c. Remember what's important

   d. Format your notes

9. Match each image to the most likely function it performs:

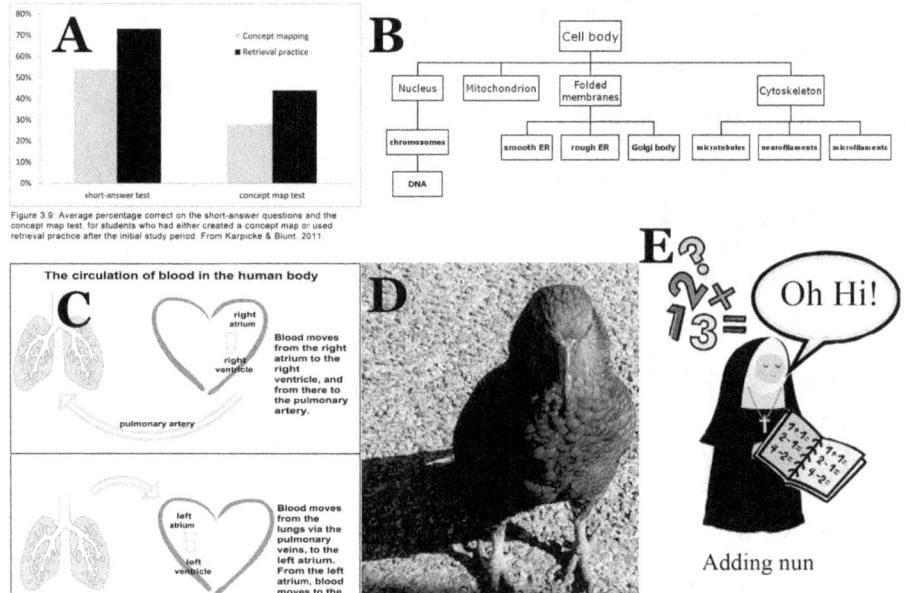

Decoration

Representation

Organization

Interpretation

Transformation

Reading

Reading is a multi-level process. While you might be able to process a narrative text on all levels simultaneously, expository texts are more difficult to process. The more difficult the text, the more useful it is to deal with it level by level.

Reading is a skill, meaning it requires practice. Skilled readers have automatized the first levels, and thus can give more attention to higher, more complex levels.

Expository texts need to be read actively. A skilled reader constructs good mental models as they read. Good mental models are vital for building expertise and overcoming the limits of working memory.

Once you have an idea of the general shape, meaning, and difficulty of the text, you can start reading it properly.

Levels of processing

One of the problems many students have is that they think that reading textbooks is the same as reading story books — you start at the beginning and work steadily through to the end. But reading is a complex skill that involves several levels of decoding and interpretation:

1. the physical features of the letters
2. the words
3. the meaningful chunks of phrases
4. the idea stated in a sentence
5. the main ideas (paragraph-level)
6. the theme (section-level).

6	A Norwegian folk tale featuring the cultural hero, the "Ash lad", as he wins the princess and the kingdom through his witty use of broken and discarded items.	Theme (section)
5	There was once a king; he had a daughter who was so wayward and willful in her speech that she always had to have the last word, and so he promised that the one who could make her hold her tongue should get the princess and half the kingdom into the bargain.	Main ideas (paragraph)
4	She always had to have the last word.	Ideas (sentence)
3	the last word	Chunks (phrase)
2	WORD	Words
1	R O D W	Letters

How effectively you read depends on your skill at each of these levels.

The first two levels involve skills that become automatic with enough practice.

The higher levels require more attention. How much attention depends on the complexity of the text. With story books, the reader generally has such a good understanding of the subject matter and how stories are structured, that even the higher levels require little conscious awareness. Expository texts require more attention.

For difficult texts, you should treat levels 3 to 6 separately, focusing on only one level at a time.

For example, here's the first paragraph of the memory gene text:

> NIH scientists have shown that a common gene variant influences memory for events in humans by altering a growth factor in the brain's memory hub. On average, people with a particular version of the gene that codes for brain derived neurotrophic factor (BDNF) performed worse on tests of episodic memory — tasks like recalling what happened yesterday. They also showed differences in activation of the

> hippocampus, a brain area known to mediate memory, and signs of decreased neuronal health and interconnections. These effects are likely traceable to limited movement and secretion of BDNF within cells, according to the study, which reveals how a gene affects the normal range of human memory, and confirms that BDNF affects human hippocampal function much as it does animals'.

Given the difficulty of this text, your first reading might focus on the technical words and phrases used: gene variant; growth factor; brain derived neurotrophic factor; episodic memory; differences in activation of the hippocampus; decreased neuronal health; secretion of BDNF within cells; human hippocampal function. There's no point trying to work out the meaning of the paragraph if you don't understand what's meant by these.

That doesn't mean you need to understand these concepts in depth, as one of the scientists would! All you need is sufficient understanding of the concepts to grasp the general ideas. So your first task is to read through the text, picking out the concepts (words and phrases) that you don't know, and searching for any explanation of them. You may find it helpful to jot down the concepts as you go, filling in your own brief explanation for any concepts as they are explained, or as other concepts shed light on them.

Once you understand the basic concepts, you can read through the text focusing on the sentences as a whole.

> NIH scientists have shown that a common gene variant influences memory for events in humans by altering a growth factor in the brain's memory hub.

> — there's a gene that affects a brain chemical that helps brain cells in the memory hub grow and increase in number

> On average, people with a particular version of the gene that codes for brain derived neurotrophic factor (BDNF) performed worse on tests of episodic memory — tasks like recalling what happened yesterday.

> — people with one version of the gene have poorer memory for events

> They also showed differences in activation of the hippocampus, a brain area known to mediate memory, and signs of decreased neuronal health and interconnections.

> — people with this gene version also show effects in the brain region concerned with memory (hippocampus)

> These effects are likely traceable to limited movement and secretion of BDNF within cells, according to the study, which reveals how a gene

> affects the normal range of human memory, and confirms that BDNF affects human hippocampal function much as it does animals'.
>
> — these effects are probably because the growth chemical is being limited in the hippocampus.

Now try and work out the main idea of the paragraph. Perhaps something like:

> People with a particular variant of the gene controlling the brain factor BDNF have reduced hippocampal function and worse memory for events.

But note that your version of the main idea should be in words that you can readily understand. If you're not yet comfortable with the technical language being used, you'll probably want to use less precise but simpler language. This will mean your sentence will be longer. This is fine, but try to incorporate some of the technical language, even if it is redundant, as a means of becoming comfortable with the new vocabulary.

As you can see, levels 4 and 5, like levels 2 and 3, can be done at much the same time. Level 6, however, requires you to have a complete understanding of the text. Once you have your complete set of main ideas, you should readily be able to work out the theme of the text:

> Human gene affects memory for events
>
> People with a particular variant of the gene controlling the brain factor BDNF have reduced hippocampal function and worse memory for events.
>
> BDNF plays a key role in memory.
>
> Two variants: "Met" variant linked to poorer episodic memory; "Val" variant more common.
>
> "Met" variant might increase risk of Alzheimer's and other disorders involving the hippocampus.
>
> Study finds those with two copies of "met" perform dramatically worse on tests of episodic memory but not on other memory tests.
>
> Two copies of "met" worse than one, but any "met" variant is associated with hippocampal dysfunction.
>
> "Met" variant less successful in distributing BDNF proteins to the synapses.
>
> Possible connection with memory problems in old age.

If you've written down your main ideas as you worked through the text, you will also, of course, have a good start on your notes!

Active reading

As should be evident by now, reading for information/instruction, unlike reading for pleasure, is a very active process.

Reading actively involves:

- thinking about what you're reading
- asking yourself questions about it
- trying to relate it to information you already know.

This is not simply about the complexity of the decoding process. It's also because of what happens when we read.

Mental models

Skilled readers build mental models as they read.

Read the paragraph in the box below.

> John was preparing for a marathon in August. After doing a few warm-up exercises, he put on his sweatshirt and began jogging. He jogged halfway around the lake without too much difficulty. Further along his route, however, John's muscles began to ache.

Don't look back! Did the word *sweatshirt* appear in the story?

Now read this:

> John was preparing for a marathon in August. After doing a few warm-up exercises, he took off his sweatshirt and began jogging. He jogged halfway around the lake without too much difficulty. Further along his route, however, John's muscles began to ache.

Did the word *sweatshirt* appear in this text?

In an experiment that had different groups read one or other of these paragraphs, those who read that John had *put on* a sweatshirt responded "yes" more quickly than those who had read that he had *taken off* his sweatshirt.

Why did this simple change make a difference?

Because reading isn't simply a matter of recognizing and finding the meaning of words. To understand a text, you build a mental model. In this case, John taking off his sweatshirt meant that the sweatshirt was no longer associated with him; your mental model discarded the sweatshirt and as he jogged through the next two sentences, John was modeled in your mind as wearing, perhaps, a T-shirt or singlet (but not a sweatshirt, anyway).

If we read about throwing a ball, the brain regions involved in throwing a ball are activated. Even metaphors such as "kicked the bucket", "saw red", "at a low point", "driving me around the bend", seem to activate the appropriate motor, spatial, or sensory regions.

When we read, we build a model in our mind of what is happening in the text.

Your skill at reading complex texts reflects:

> - the models you build, and
> - your skill at building them.

The mental model you build:

> - underlies your understanding of the text
> - is constructed from both the information in the text and your existing knowledge
> - provides a retrieval structure that you use when you later try to remember what you read.

Your existing knowledge therefore becomes more and more important, the more difficult the text is.

The difficulty of a text directly reflects the coherence of the text. A coherent text is one from which you can construct a single mental model.

In other words, how difficult you find a text depends on the mental model you can construct from it. The more difficult the text, the harder it will be to build a single mental model — unless you have a good amount of relevant knowledge that helps fill in the gaps in the text.

Why are mental models so important?

Because our working memory is so limited.

Think about reading. Even with a simple text, you need to remember what's gone before to make sense of it. But working memory is so small — how can we hold all the information we need to make sense of what we're reading? Shouldn't there be constant delays as we access needed information from long-term memory? But (unless we're finding the text difficult) there aren't.

The answer lies in **long-term working memory**, which can be thought of as a retrieval structure allowing a network of linked concepts to remain readily available.

Think about reading a difficult text in a subject you know well. Compare this to studying a difficult text in a subject you don't know well. In the second case, you probably have to:

> backtrack

> check earlier statements

> try to remember what was said before

> try to relate what you are reading to things you already know.

When you know the subject well, you don't need to do that; rather, you seem to have a vastly expanded amount of readily accessible relevant information, from the text itself and from your long-term memory. This is because the strong and well-connected networks you have in long-term memory can be accessed, and kept in readiness, as single chunks.

Building expertise is a means of increasing your functional working memory capacity.

Building expertise is, in part, a matter of learning to build good mental models.

This isn't only about knowing a subject very well. There are also general skills and areas of knowledge that help you build expertise in any subject. One vital skill is general expertise in reading expository texts.

Reading expository texts

As I've said, most texts fall into one of two kinds:

> those that tell stories (narrative texts)

- those that tell you about stuff (expository texts).

Narrative texts are easy, because:
- the structure is more straightforward
- we're experts in story structure — because we've been reading and hearing stories since we were small children
- we're 'naturally' attuned to story structure (perhaps).

Expository texts are much harder, because:
- there are several different text structures that can be used
- the text structures are more complex
- we don't have the same amount of experience in dealing with these text structures
- vocabulary used is often technical.

How narrative & expository texts differ

Narrative & expository texts require different mental models

A story mental model involves:
- characters
- their emotional states
- setting
- action
- a sequence of events that is reasonably predictable.

An expository mental model may involve:
- the components of a system
- their relationships
- events and processes that occur during the working of the system
- uses of the system.

Or other elements — because expository text can use a number of different text structures.

Narrative & expository texts use different cause-&-effect patterns

- In stories (as in everyday life), cause-&-effect is understood in terms of *goal* structures (hence superstitions — we attribute purposefulness to almost everything!). People and animals do things because they want something.

- In scientific text, cause-&-effect usually needs to be understood in terms of *logical* structures. Things happen because of built-in consequences (the leaf falls because, having become detached, it can no longer resist the effects of gravity; we don't say it falls because it 'wants' to fall!).

Expository texts are remembered much more easily when:

- they are human-focused
- they can be understood in terms of goal structures.

The more a text is like a story, the easier it will be to process and remember.

Narrative & expository texts have different inference patterns

- Stories have a lot of forward / predictive inferences ("the woman fell out of the 10th story window … Her orphaned daughters sued" — the reader easily infers that the woman is dead; it doesn't have to be explicitly stated).

- Expository texts typically involve a lot of backward inferences ("Chlorine compounds do not react with other substances. They make good propellants" — from this, one can surmise that propellants don't react with other substances; that is, the reader has to look backward, to the previous statement, to make the necessary inference).

Inference making is crucial to comprehension and the construction of your mental model, because a text never explains (*can't* explain) every single word and detail, every logical or causal connection.

Moreover, research has shown that having gaps in the text (**coherence gaps**) is important for learning — if the writer spells out too much, the reader may fool themselves into thinking they have grasped the material without having properly processed it.

But there are some problems with coherence gaps:

- ➤ The 'right' amount of coherence gaps varies depending on the expertise of the reader.
- ➤ Students often don't notice that there are gaps in the text which require inferences to be made.
- ➤ In particular, readers frequently don't notice that something they're reading is, or appears to be, inconsistent with something they already believe, or have read.

This goes back to the limitations of working memory — limitations, remember, that can be overcome through expertise in the topic.

Approaching expository text as a novice

- ➤ Build your understanding of text structure (both knowledge of the various kinds of expository structures, and knowledge of the cues that indicate what type of structure a text has).
- ➤ Ask questions of the text, searching it for the answers. Good questions help you make connections, both within the text, and between the text and your existing knowledge.
- ➤ Seek to check:
  - ★ your understanding of the words and sentences
  - ★ how well the information in the text agrees with what you already know
  - ★ how well the information in the text agrees with other information in the text
  - ★ how well the statements in the text lead from one to the other
  - ★ how well the statements fit together and support the main theme
  - ★ how clear and complete the text is.

Active reading strategies

Active reading strategies you can use on expository text:

Search strategies:

- jumping forward or backward in the text to find particular information
- skimming through the text for particular information
- anticipating information that might be covered in the text, and hunting for it
- looking out for cues

Clarifying strategies:

- backtracking for clarification
- attending to figures and tables
- re-stating text in your own words (summarizing)
- generating questions

Elaborating strategies:

- moving back and forth between different parts (perhaps between a table and the text) to integrate them
- thinking of analogies or examples
- making inferences; drawing conclusions
- constructing appropriate images (directed **imagery**)

Evaluating strategies:

- analyzing the text structure
- assessing the difficulty of the text — i.e., how hard it is (for you) to understand (comprehension monitoring strategies include periodic review; self-interrogation)
- noting what information you already knew
- evaluating the relevance of the information to your own goals
- evaluating the quality of the information — its internal consistency; how compatible it is with information you already know.

Main Points

Understanding expository text is a skill that must be learned (it's not 'natural' in the way story texts are).

Expository text is easier to understand the more closely it resembles narrative text, with a focus on goals and human agents.

How well the text is understood depends on the amount and extent of the coherence gaps in the text relative to the skills and domain knowledge of the reader. So the less you know of the subject, the fewer and smaller coherence gaps you can cope with.

The more difficult the text, the more you should break down the reading process.

Read expository texts actively — treat the text as a partner in your dialog: ask it questions; challenge its meaning.

Consciously build your mental model:

> - use the text structure as a guide to how you should structure your model
> - make connections between different statements in the text
> - make connections to information you already have
> - if you find holes in the text — places where information that would make sense of what's said has apparently been omitted — recognize these as coherence gaps which the reader is obliged to fill from their own knowledge. If you don't have the knowledge to fill them, try and find it elsewhere.

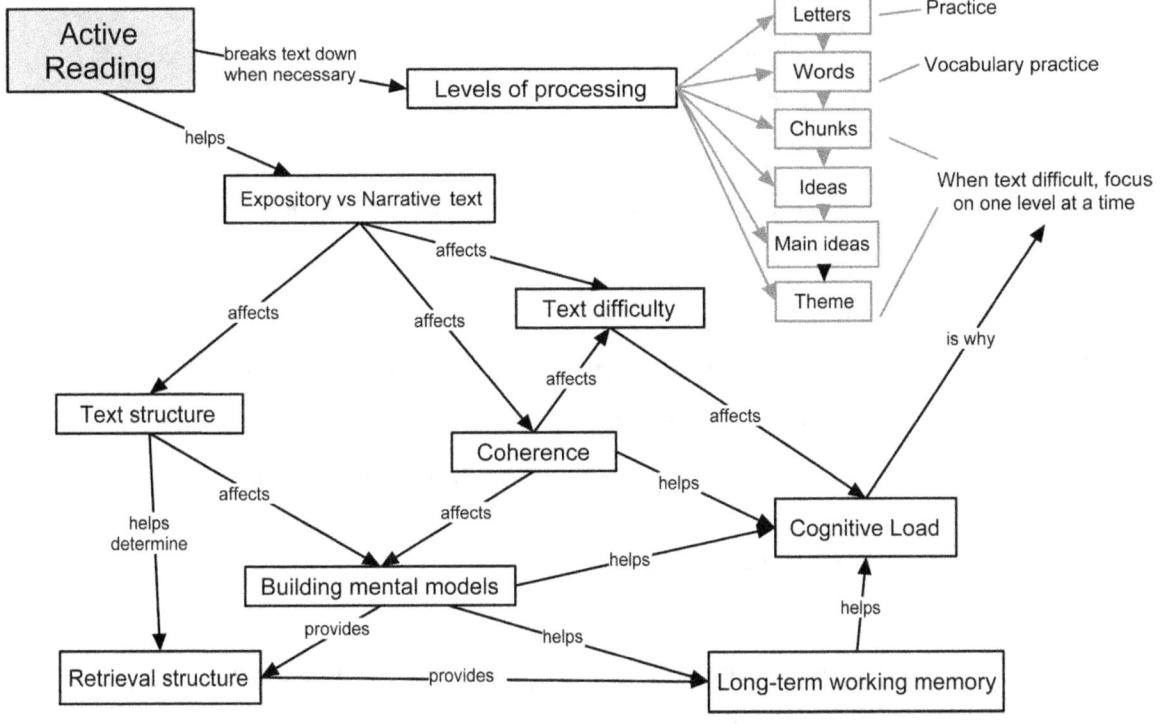

Review Questions

1. Expository texts are more difficult to understand and remember than narrative texts because:

    a. We have less experience in reading the types of text structure found in expository texts

    b. The structure of expository texts is more varied

    c. We have less experience in reading expository texts

    d. The content of expository texts tends to be more difficult

Reading 91

2. Active reading strategies include:

   a. Asking yourself questions

   b. Backtracking

   c. Trying to remember every word

   d. Skimming

   e. Evaluating the quality and relevance of the content

3. A coherence gap is:

   a. A measure of how far you are from understanding the text

   b. Garbled speech or text

   c. A place in the text where omitted information requires the reader to make an inference

   d. A missing phrase or sentence

4. Which of these are NOT levels in the process of reading:

   a. Identifying the general theme of the section

   b. Understanding the meaning of each sentence

   c. Decoding the shape of the letters

   d. Building a mental model

   e. Understanding each word

   f. Identifying the main idea of each paragraph

   g. Understanding each phrase

Taking Notes

Note-taking is a vital skill for learning, because it's when you make the information your own. It's a visible form of encoding. That's why using your own words, and re-organizing and transforming the information, are so critical for effective note-taking.

If you treat note-taking simply as a means of taking a record of information you want to remember, then you are losing more than half of its value.

Notetaking makes information meaningful

Why take notes?

- To give you a record of the information you want to remember.
- This is NOT the main benefit of notetaking from texts (as opposed to lectures).
- To take control of your brain's editing and organizing process.
- This IS the main cognitive benefit of notetaking, and is the reason notetaking is only effective to the extent that you **paraphrase, organize and make sense of the information.**

How difficult notetaking is depends on slightly different qualities for information you hear compared with material you read:

- Audio material (e.g., lectures) depends on:
    - ★ how fast the lecturer speaks
    - ★ how organized the lecturer is
    - ★ how dense the information content is.
- Written material (e.g., textbooks) depends on:
    - ★ how dense the text is
    - ★ how complex the text is
    - ★ how organized the text is
    - ★ the 'tone' of the text.

Why do some students find it harder to cope with these problems?

- poor strategies
- little skill in using good strategies
- ignorance of the topic and relevant background knowledge
- reduced working memory capacity.

The first and most crucial step in effective notetaking is selecting what information is important.

Selection becomes more difficult:

- **the more dense the material**

- **the less organized the material**
- **the faster the information is presented.**

The more you know about a topic, the easier it is to pick out what's important.

This is why most students take lots of notes when they are new to the subject and become more selective with time. It's also why it helps to read relevant material before a lecture — and why this is especially helpful when the lecturer is fast and/or disorganized.

WMC impacts selection skill by affecting how much information you can work on at one time.

You can help this by automating component skills:

- Skills become automatic through being practiced sufficiently.
- Automated skills put very little burden on working memory.

One selection skill is **cue recognition** (recognizing the cues that indicate what information teachers or writers regard as important).

Another is learning to recognize conditions that make selection difficult.

Highlighting

You can pick out what's important as you read, and a common strategy is to highlight such information, either by underlining or by using a colored marker.

This is not a very effective strategy, but does have its place when done right.

Here's how not to do it:

> At the height of the Ice Age, <u>between 34,000 and 30,000 B.C., much of the world's water was locked up in vast continental ice sheets. As a result, the Bering Sea was hundreds of meters below its current level, and a land bridge, known as Beringia, emerged between Asia and North America. At its peak, Beringia is thought to have been some 1,500 kilometers wide.</u> A moist and treeless tundra, it was covered with grasses and plant life, attracting the large animals that early humans hunted for their survival.
>
> The first people to reach North America almost certainly did so without knowing they had crossed into a new continent. They would have been following game, as their ancestors had for thousands of years, along the Siberian coast and then across the land bridge.

Once in Alaska, it would take these first North Americans thousands of years more to work their way through the openings in great glaciers south to what is now the United States. Evidence of early life in North America continues to be found. Little of it, however, can be reliably dated before 12,000 B.C.; a recent discovery of a hunting lookout in northern Alaska, for example, may date from almost that time. So too may the finely crafted spear points and items found near Clovis, New Mexico.

Similar artifacts have been found at sites throughout North and South America, indicating that life was probably already well established in much of the Western Hemisphere by some time prior to 10,000 B.C. Around that time the mammoth began to die out and the bison took its place as a principal source of food and hides for these early North Americans. Over time, as more and more species of large game vanished whether from overhunting or natural causes plants, berries, and seeds became an increasingly important part of the early American diet. Gradually, foraging and the first attempts at primitive agriculture appeared. Native Americans in what is now central Mexico led the way, cultivating corn, squash, and beans, perhaps as early as 8,000 B.C. Slowly, this knowledge spread northward.

By 3,000 B.C., a primitive type of corn was being grown in the river valleys of New Mexico and Arizona. Then the first signs of irrigation began to appear, and, by 300 B.C., signs of early village life.

By the first centuries A.D., the Hohokam were living in settlements near what is now Phoenix, Arizona, where they built ball courts and pyramid like mounds reminiscent of those found in Mexico, as well as a canal and irrigation system.

This level of highlighting is common — typically, students highlight about 70-80% of a text.

Think about this: by highlighting so much of the text, you haven't greatly reduced the content to be learned. Moreover, you have failed in your task of selecting only the most important information.

Selecting only the most important information does *not* mean that you're making a choice to only remember the few bits you have chosen. Selecting enables you to focus on the truly important information, which will provide anchors to other details. So don't panic that you're being too restrictive. You need to build a strong framework before you can decorate it with all the little details you hope to remember. Concentrate on that.

The rule of thumb for effective highlighting is about 10% of the text. For example, the text on early Americans is some 39 lines, so you would aim to highlight around 4 lines in total. Let's try cutting it back:

At the height of the Ice Age, between 34,000 and 30,000 B.C., much of the world's water was locked up in vast continental ice sheets. As a result, the Bering Sea was hundreds of meters below its current level, and a land bridge, known as Beringia, emerged between Asia and North America. At its peak, Beringia is thought to have been some 1,500 kilometers wide. A moist and treeless tundra, it was covered with grasses and plant life, attracting the large animals that early humans hunted for their survival.

The first people to reach North America almost certainly did so without knowing they had crossed into a new continent. They would have been following game, as their ancestors had for thousands of years, along the Siberian coast and then across the land bridge.

Once in Alaska, it would take these first North Americans thousands of years more to work their way through the openings in great glaciers south to what is now the United States. Evidence of early life in North America continues to be found. Little of it, however, can be reliably dated before 12,000 B.C.; a recent discovery of a hunting lookout in northern Alaska, for example, may date from almost that time. So too may the finely crafted spear points and items found near Clovis, New Mexico.

Similar artifacts have been found at sites throughout North and South America, indicating that life was probably already well established in much of the Western Hemisphere by some time prior to 10,000 B.C. Around that time the mammoth began to die out and the bison took its place as a principal source of food and hides for these early North Americans. Over time, as more and more species of large game vanished whether from overhunting or natural causes plants, berries, and seeds became an increasingly important part of the early American diet. Gradually, foraging and the first attempts at primitive agriculture appeared. Native Americans in what is now central Mexico led the way, cultivating corn, squash, and beans, perhaps as early as 8,000 B.C. Slowly, this knowledge spread northward.

By 3,000 B.C., a primitive type of corn was being grown in the river valleys of New Mexico and Arizona. Then the first signs of irrigation began to appear, and, by 300 B.C., signs of early village life.

By the first centuries A.D., the Hohokam were living in settlements near what is now Phoenix, Arizona, where they built ball courts and pyramid like mounds reminiscent of those found in Mexico, as well as a canal and irrigation system.

Still too much — and here we come to a major reason why highlighting is not an effective means of taking notes. You have to use the exact words given in the text.

Here's the first paragraph again:

> At the height of the Ice Age, between <u>34,000 and 30,000 B.C.</u>, much of the <u>world's water was locked up in vast continental ice sheets</u>. As a result, the <u>Bering Sea was hundreds of meters below its current level</u>, and <u>a land bridge, known as Beringia, emerged between Asia and North America</u>. At its peak, Beringia is thought to have been <u>some 1,500 kilometers wide</u>. A moist and treeless tundra, it was covered with grasses and plant life, attracting the large animals that early humans hunted for their survival.

What we really want from this is:

- 34,000-30,000 BC
- water locked up in ice → sea lower, exposing land bridge between Asia and America (Beringia)
- 1,500 miles wide

In the former, about 40 words are highlighted. The bullet points say the same thing in just over half the words (24 words).

Rather than helping you pick out important information, the main benefit of highlighting is to help you focus your attention.

Highlighting is a stepping-stone strategy, most useful when:

- you are new to the topic, or
- you are building your selection skills, or
- you have a low WMC.

Most especially, when all these are true!

However, highlighting can be useful as a simple strategy on its own:

- when you can easily highlight no more than 10% of the text, and
- the highlighted text is concise and understandable enough without any paraphrasing.

This is most likely to occur when all these are true:

> you have some expertise in the topic

> you have well-developed selection skills

> the text is simple.

Whether highlighting is an effective strategy also depends on your goals.

For example, if all you needed to get out of the text above was the answer to the question "When did people first arrive in North America?", it would be quick and easy to highlight the relevant information. If, however, your goal was to answer the question: "What evidence do we have with respect to the early settlement of the American continent?", then the relevant information is both greater and more inferential.

The number of important points in a text isn't, therefore, simply a property of the text itself, but also depends on what you want from it.

What highlighting does

Highlighting acts to direct and focus attention.

Highlighting encourages you to spend more time with the material.

Highlighting improves memory for highlighted information — but at the expense of the rest of the text.

Highlighting doesn't help you remember highlighted details any better if the text is short.

Guidelines to using highlighting effectively

Restrict your highlighting to no more than 10% of the text — i.e., one line in ten.

Select those details that you want to remember, and *don't think you will* (don't highlight facts simply because they are important, if you already know them, or believe you will remember them without help).

In information-dense text, or where the text is difficult, use highlighting only in conjunction with other strategies.

Don't use it if your time is limited.

Only use one color or method of highlighting (multiple colors add to cognitive load and are more likely to harm learning than help it).

Creating summaries

Summarizing effectively depends entirely on your skill at distinguishing important information from less important.

Estimates suggest perhaps half of all college students can't summarize effectively.

Most commonly, they'll use the 'copy-delete strategy': copy out sentences they think are important, leaving out the ones they don't regard as important.

This has the same fatal flaw as highlighting: main ideas are rarely completely contained in a single sentence.

Your goal is to tease out all the important information contained in related sentences and combine them into a new sentence or sentences, that are as brief as can be without omitting important information.

A good summary:

> - is short
> - contains only, and all, the most important information
> - is in your own words.

Here are five "rules" beginners are told to use to get started:

1. delete trivial material
2. delete redundant material
3. replace lists of items with a superordinate term
4. select a topic sentence
5. if no topic sentence given, create one.

But these 'trainer wheels' are of limited value. Most students will know to ignore trivial and redundant information, and to collapse items that belong together into one overarching category. The most difficult part of summarizing effectively (and what this set of 'rules' gives no help with) is the construction of sentences that sum up a paragraph.

In fact, while it's helpful to bear these steps in mind, I think it stands in the way of good summarizing to follow them in a mechanistic fashion. If you are an absolute beginner, you might find it helpful to approach a few texts in this way, but it's probably counter-productive to do this for long.

Here are my rules:

DON'T:

start at the beginning and steadily work through the text, discarding unimportant details as you go, identifying or constructing topic sentences paragraph by paragraph.

DO:

1. Get a feel for the text before you read it, either by:
   - ★ reading any summaries provided, or (if not provided)
   - ★ studying headings, or (if headings not frequent and descriptive)
   - ★ skimming the text.
2. Put in your own words what you think the main idea of the text is.
3. Identify the text structure. Use this as a framework for your notes.
4. Read the whole text before taking any notes.
5. Check your main idea. You may want to amend or re-word it.
6. Identify the supporting ideas: group together paragraphs on the same topic; write in your own words what the idea is.
7. Use the text structure as a framework; see how well these ideas fit into it.
8. Take a short break.
9. Without looking at the text, write down the main idea, supporting ideas, and any important details you remember.
10. Look back at the text. Amend anything you got wrong; add any important information you omitted.

("Main idea" and "supporting ideas" may be topics and subtopics, topics and categories, topics and steps, etc, depending on the text structure.)

Let's see this with one of our example texts. Here are the headings and the provided overview:

> The Relationship Of Ozone And Ultraviolet Radiation: Why Is Ozone So Important?
>
> Most of the ozone in our atmosphere is held in the stratosphere, in what's called the ozone layer, which protects us from harmful ultraviolet

radiation. However, some is found closer to the surface, in the troposphere, where it is a pollutant. Ultraviolet waves are dangerous because they're energetic enough to damage DNA, but the shortest waves are blocked by the ozone layer, and the longest are not so damaging, so the main problem are those in the middle (UV-b). Time and season affect how much of this radiation is absorbed by the ozone layer, because the angle of the sun affects how long the radiation takes to pass through it.

2.1 Ozone and the Ozone Layer

2.2 Solar Radiation

2.3 Solar Fluxes

2.4 UV Radiation and the Screening Action by Ozone

Articulate the main idea:

> ➢ Ozone when high in the atmosphere protects living organisms from the damage ultraviolet waves can inflict.

Skim the text looking for signal words:

> ➢ because; types; because; in contrast to; many; classified; therefore; because; because; because; in addition; because; because; classify; so that; because; consequently

From this, and the theme of the text (the main idea helps with that), identify the text structure:

> ➢ Cause-&-effect, plus some classification

Check your main idea: do you need to amend it in light of your better understanding?

Identify the supporting ideas from each paragraph or set of related paragraphs:

1. Ozone is important because it shields the surface from harmful ultraviolet radiation.

2. The stratosphere holds 90% of the ozone in our atmosphere (the ozone layer); the troposphere holds 10%.

3. Ozone absorbs UV radiation. Of the three different types of ultraviolet (UV) radiation, the shortest (UV-c) is entirely screened out by the ozone layer, while the longest (UV-a) is not so damaging, so the main problem is UV-b.

4. The ozone layer protects us. Because ozone is harmful to breathe in, tropospheric ozone is a pollutant. It is found in high concentrations in smog.

5. The Sun produces radiation at many different wavelengths: the electromagnetic spectrum (EM). Wavelength is a measure of how energetic the radiation is.

6. Radiation with wavelengths shorter than those of violet light (at the short end of the visible spectrum) is called ultraviolet radiation.

7. UV waves are dangerous because they're energetic enough to break the bonds of DNA molecules.

8. Solar flux = the amount of solar energy in watts falling perpendicularly on a surface one square centimeter.

9. The action spectrum measures the relative effectiveness of radiation in generating a certain biological response (such as sunburn) over a range of wavelengths. Because ozone is most protective on the most dangerous wavelengths, a 10% decrease in ozone would increase the amount of DNA-damaging UV by about 22%.

10. expands on para 3

11. Time and season affect how much UV radiation is absorbed by ozone because the angle of the sun affects how long the radiation takes to pass through the atmosphere (the path is shorter when the sun is directly overhead, so the radiation meets fewer ozone molecules).

Now try and fit these ideas into a cause-&-effect structure:

Topic: Ozone & UV radiation

Cause: the sun emits UV radiation

Effect: UV radiation can damage DNA

Cause: ozone absorbs UV radiation

Cause: ozone is concentrated in the stratosphere

Effect: stratospheric ozone (ozone layer) blocks out most UV radiation

Cause: some ozone is found lower to the ground, in the troposphere

Effect: breathing it in is unhealthy, particularly to eyes and lungs

Cause: reduction in the ozone layer

Effect: more genetic damage to living thing

Ultraviolet radiation comes in 3 types:

- ★ UV-c: shortest wavelength; entirely screened out by the ozone layer
- ★ UV-a: longest wavelength; not so damaging to living organisms
- ★ UV-b: wavelength between UV-a and UV-c; the most dangerous

Cause: ozone is most protective on the most dangerous wavelengths

Effect: a 10% decrease in ozone would increase the amount of DNA-damaging UV by about 22%.

Cause: the angle of the sun affects how long the radiation takes to pass through the atmosphere

Effect: time and season affect how much UV radiation is absorbed by ozone

Measurement:

> Solar flux = the amount of solar energy in watts falling perpendicularly on a surface one square centimeter.

> The action spectrum measures the relative effectiveness of radiation in generating a certain biological response (such as sunburn) over a range of wavelengths.

Check how well the framework fits; whether any important information has been omitted.

How to deal with different types of text structure

Text structures are basically frameworks. To create a good mental model that will help you remember the text and produce good notes, it helps to use the same framework (i.e. structure) the text is built on.

Unfortunately, the memorability and effectiveness of a text structure is in direct opposition to that structure's popularity! The most common structures (description; collection) are the least effective for learning and recall; the least common (refutation) is the most effective. You may therefore want to be flexible on this issue. Use the text structure as a guide, but don't feel bound by it if it doesn't work well for your purposes.

So, for example, using the sequence structure, you might produce a summary for the blood flow text that is something like this:

Topic: blood flow in human body

Step: pulmonary circulation — movement from heart to lungs via **pulmonary artery**, and lungs to heart via **pulmonary veins**

Step: coronary circulation — movement within the 4 chambers of the heart: **right atrium → right ventricle →** (pulmonary artery / lungs / pulmonary veins) **→ left atrium → left ventricle →** (aorta)

Step: systemic circulation — movement in the rest of the body: **aorta → arteries → capillaries → veins**

This emphasizes the movement of the blood. You might, however, be more interested in the why and how. Accordingly, you might want to try using a cause-&-effect structure:

Cause: the heart pumps blood

Effect: blood flows in spurts

Cause: the right atrium fills with blood

Effect: it contracts, pushing the blood into the pulmonary artery

Effect: blood is carried to the lungs

Effect: the blood binds the incoming oxygen and loses the carbon dioxide it carried, and is pushed into the pulmonary veins

Effect: the blood is carried back to the heart, filling the left atrium

Effect: the left atrium contracts, pushing the blood into the left ventricle

Effect: the left ventricle contracts, pushing the blood into the aorta

Effect: the blood is pushed out into the other arteries, carrying the fresh blood throughout the body

Effect: the blood reaches the capillaries, where the oxygen and nutrients are released

Effect: the blood reaches the veins

Effect: the blood returns to the heart

Spelling this out makes it clear that a sequence structure is more appropriate, but also highlights valuable details about how it all works. You could either incorporate those details into your sequence structure summary, or choose to collect those details into a separate item.

> Step: pulmonary circulation — movement from heart to lungs via **pulmonary artery**, and lungs to heart via **pulmonary veins** — within the lungs, the blood binds the incoming oxygen and loses the carbon dioxide it carried
>
> Step: coronary circulation — movement within the 4 chambers of the heart: **right atrium → right ventricle →** (pulmonary artery / lungs / pulmonary veins) → left atrium → left ventricle →(aorta) — each chamber contracts when filled with blood, pushing the blood out
>
> Step: systemic circulation — movement in the rest of the body: **aorta → arteries → capillaries → veins** — oxygen and nutrients are released are released in the capillaries, before heading back to the heart via the veins

Here are some more hints and examples for dealing with different types of text structure.

Description / Generalization

- ➢ identify the main idea
- ➢ list and define the key words
- ➢ restate the main idea in your own words
- ➢ look for evidence to support the main idea
   - ★ what kind of support is there for the main idea?
   - ★ are there examples, illustrations?
   - ★ do they extend or clarify the main idea?
   - ★ can you think of any supporting evidence from your own knowledge?

Example (using *The role of consolidation in memory*):

Main idea: memories aren't as stable and fixed as we think

Keywords:

>consolidation: the process of 'fixing' a new memory
>
>synapse: some structure of brain cells that's involved in encoding memories
>
>labile: unstable; sensitive to disruption

reconstructive: retrieval is a matter of rebuilding the memory rather than simply replaying it

reconsolidation: memories may be consolidated anew each time they're retrieved

hippocampus: the brain region most involved in consolidation, perhaps only for the first few years; involved in episodic learning

medial temporal lobe: a brain region that contains the hippocampus and entorhinal cortex, among others

entorhinal cortex: the brain region involved in long-term consolidation; perhaps involved in incremental learning

incremental learning: learning that requires repeated experiences

episodic learning: memories that require only a single experience, such as the memories of events

Supporting details:

the formation of new synapses requires repeated stimulation of the cell

various neurological processes (such as glutamate release and protein synthesis) are needed to stabilize a new memory

it now appears stable memories become labile again when reactivated

reconsolidation may involve creating a new representation or changing the old one

damage to the entorhinal cortex is seen early in the development of Alzheimer's

memories may stay in the hippocampus for some years, but then move on to the entorhinal cortex

Collection / Enumeration

- name the topic
- identify the subtopics
- organize and list the details within each subtopic, in your own words
- search for a connective thread that ties the items together and might help you remember them (it may have to be mnemonic — this structure is the one most likely to require mnemonic assistance).

Example (using *Introducing brain cells*):

- ➤ Neurons: much less common than glia; do most of the information-processing work
  - ★ Neuron parts:
    - Soma/cell body: roughly spherical
      - nucleus:
        - chromosomes:
          - DNA
      - folded membranes:
        - rough endoplasmic reticulum:
          - ribosomes: involved in protein synthesis
        - smooth endoplasmic reticulum
        - Golgi apparatus: involved in post-assembly protein processing
      - mitochondria: makes energy for cell
    - Neurites:
      - axon: usually only one; uniform thickness; much longer than dendrites, channel for output
      - dendrites: many; tapers; very short; incoming receiver
        - synapse
    - Cytoskeleton: controls cell shape
      - microtubules: biggest; made from polymers
      - neurofilaments: made from single long protein molecules; very strong and stable; can go wrong
      - microfilaments: smallest; made from polymers
- ➤ Neuron types
  - ★ unipolar; bipolar; multipolar
  - ★ pyramidal; stellate
  - ★ spiny; aspinous
  - ★ primary sensory; motor; interneurons
  - ★ projection; local circuit
  - ★ neurotransmitters

- Glia
  - Astrocytes
  - Oligodendroglia
  - Schwann cells

Cause & Effect
- identify the topic
- identify each cause
- identify and briefly describe each effect/consequence
- pay particular attention to the connections between the causes
- pay particular attention to any accompanying illustration.

Example given earlier.

Sequence
- identify the topic
- name each step and outline the details within each
- briefly discuss what's different from one step to another
- pay particular attention to the causal connections linking the steps together, as these are especially memorable
- pay particular attention to any accompanying illustration.

Example given earlier.

Classification
- identify the topic
- name the categories and briefly describe the similarities between the category members
- briefly discuss what makes the categories different from each other
- pay particular attention to similarities and differences that are meaningful to you and will be memorable.

Example (using *How blood flows*):

    Circulatory system: pulmonary; coronary; systemic.

    Pulmonary circulation: between heart and lungs

    Coronary circulation: within the heart.

    Systemic circulation: rest of the body.

    Aorta → smaller arteries → capillaries → veins

    Arteries carry fresh, oxygenated blood

    Capillaries release the oxygen

    Veins carry the de-oxygenated blood

    The heart powers the system.

    Blood enters the heart via two large veins, and leaves via the main artery.

    Carbon dioxide is removed and replaced with oxygen in the lungs.

Compare / contrast

- identify the topic
- name the items being compared
- describe their differences &/or similarities
- pay particular attention to similarities and differences that are meaningful to you and will be memorable.

Example (using *Human gene affects memory*):

    Topic: A gene that affects memory

    Items compared: Val variant of BDNF gene; Met variant of BDNF gene

    Differences: differ in one amino acid; met version has the amino acid methionine where the val version has valine; met version is linked to poorer episodic memory; met version is linked to lower levels of a marker signaling cell health and number of synapses; met version linked to clumping of BDNF inside cell body, while val version is linked to BDNF spreading out and into dendrites

    Similarities: variants of the same gene; gene codes for brain-derived neurotrophic factor; affects episodic memory; doesn't affect other types of memory; affects activation in the hippocampus

Problem

- ➢ identify who has the problem and what the problem is
- ➢ briefly describe the reasons for the problem (if appropriate)
- ➢ name the action taken to solve the problem
- ➢ describe what happened as a result of the action taken.

Example:

Problem: 51% of British children aged 7-11 unable to swim 25 meters

Reason: lack of funding; cost of transport; limited access to a pool

Solution: program to teach trainee teachers how to give swimming instruction

What happened: n/a (solution proposed only)

Refutation

- ➢ describe the misconception
- ➢ explain why it's common
- ➢ describe the actual facts
- ➢ discuss the evidence for them

Example:

Misconception: some people are "left-brained" and others are "right-brained"; left-brained people are logical and analytical while right-brained people are artistic and creative.

Why it's common: people like labels; some people are indeed more analytical than others, while some are more creative.

Fact: Among right-handers, language is usually mainly processed in the left hemisphere. (Evidence: Brain studies)

Fact: People's minds vary on a number of dimensions, including working memory capacity, imagery ability, and anxiety level. These dimensions can interact. (Evidence: Cognitive tests)

Fact: All cognitive tasks activate areas all over the brain. (Evidence: Brain studies)

Fact: Specific brain areas can become larger or more richly connected through use. (Evidence: London taxi drivers who have passed the "Knowledge" test have been found to have a larger right posterior hippocampus (an area involved in spatial navigation)).

Summarizing isn't just about the notes you take

Summarizing is a process. Summarizing involves thinking about the text and working out what it means and what matters.

A lot of that thinking doesn't need to be noted down.

The length of a summary depends on your understanding:

> ➤ The less you understand, the fuller your notes will tend to be — because you can't tell what's important and what's not.

> ➤ The more you understand, the shorter your notes will be — because you've captured the gist, you've recorded only what's important.

(Of course, this generalization only applies to diligent students! Your notes may be short when you don't understand because you have given up on understanding the text.)

As the use of text structures suggests, summarizing isn't only about creating topical summaries (linear summaries of the main points). You can also create graphic summaries, and these can be in a variety of formats.

Outlines and Graphic Organizers

```
            Branches of Government (U.S.A.)
  I.  Executive Branch
      A.  Represented by:   President
      B.  Powers:           Can recommend legislation;
                            veto legislation; appoint judges
      C.  Length of term:   4 years; maximum term 8 years
  II. Legislative Branch
      A.  Represented by:   Congress
      B.  Powers:           Can enact legislation; override veto; reject
                            & impeach judges; impeach President
      C.  Length of term:   2 years (House of Representatives) or 6
                            years (Senate); no maximum term
  III. Judicial Branch
      A.  Represented by:   Supreme Court and other federal courts
      B.  Powers:           Can declare legislation unconstitutional
      C.  Length of term:   Life
```

In an outline, topics are listed with their subtopics in a tabular format, as in this table (above). Graphic organizers show the same sort of information, but in a more visual format, like this tree diagram below.

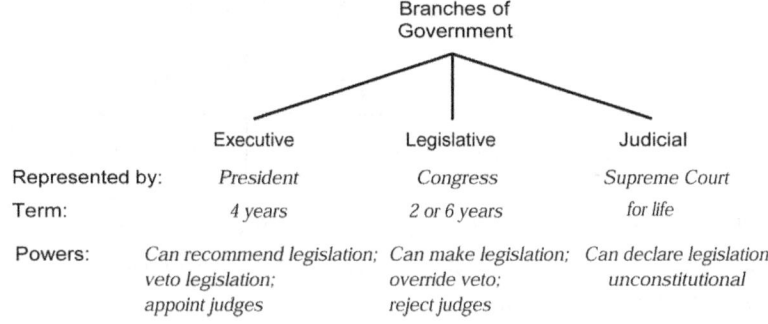

> Compare the tabular outline and the tabular matrix.

Or this matrix:

	Executive Branch	Legislative Branch	Judicial Branch
Represented by:	President	Congress	Supreme Court
Length of term:	4 years	2 or 6 years	Life
Powers:	Can recommend legislation; veto legislation; appoint judges	Can enact legislation; override veto; reject & impeach judges; impeach President	Can declare legislation unconstitutional

Both outlines and graphic organizers are very helpful for hierarchical information (and for hierarchical information *only*).

But graphic organizers

> ➤ are helpful for showing relationships *between* information clusters (you can readily compare the different terms for each of the branches, for example)

While outlines

> ➤ are useful for showing hierarchical information and relationships within clusters, but not for displaying relationships between clusters

Trying to put suitable information in this format is also an excellent means of revealing the gaps in your information. For example, take the text on the early Americans. The details about the various pre-Columbian peoples of the U.S. would seem to lend themselves to a tabular form, allowing us to see the changes in time and place. However, if we try and tabulate these details from this text, we end up with something like this table on the next page.

Taking Notes   113

Adenans	600 B.C. — ?	?
Hohokam	first centuries A.D	Arizona etc
Hopewellians	? — 500 A.D.	southern Ohio etc
Mississippians / Temple Mound culture	Peak: early 12th C	Illinois etc
Anasazi	900 A.D. — ?	Southwest U.S.

This makes clear that if we want to have a clear timeline, and picture of the territories they covered, we'll have to search elsewhere. A quick look at Wikipedia suggests some rough dates:

Adenans	1000—200 B.C.E.
Hohokam	1—1450 C.E.
Hopewellians	200—500 C.E.
Mississippians / Temple Mound culture	800—1500 C.E.
Anasazi	900—1150 C.E.

However, the territorial information is less able to be given a clear label (because of the mismatch between those territories and modern regions), and another type of format is better for that.

More graphic organizers

Graphic organizers are basically diagrams that show the relationships between pieces of information. Different types of text structure lend themselves to different formats (note that these are suggestions only, not hard-and-fast rules!):

> **Description/Generalization**: topical summary; spider map; concept map; mind map

> **Collection/Enumeration**: outline; graphic organizer (tree; matrix; network diagram)

> **Cause-&-effect**: multimedia summary; loop network diagram; fishbone diagram

- **Sequence**: multimedia summary; timeline; linear or loop network
- **Classification**: outline; graphic organizer (tree; matrix; tree-like network diagram)
- **Comparison / contrast**: graphic organizer (Venn diagram; T-chart; matrix; network diagram)
- **Problem**: graphic organizer (fishbone diagram)
- **Refutational**: topical summary; graphic organizer (T-chart; refutation map)

Again, I don't want to suggest that these are the only strategies appropriate for these structures, or that they will always be appropriate for that structure.

You can see some examples of these below and on the following pages (concept and mind maps are covered in the next section).

Network diagrams

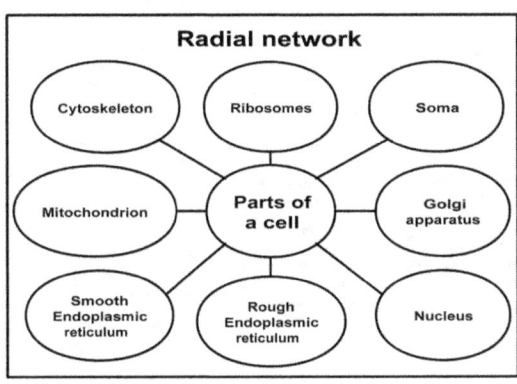

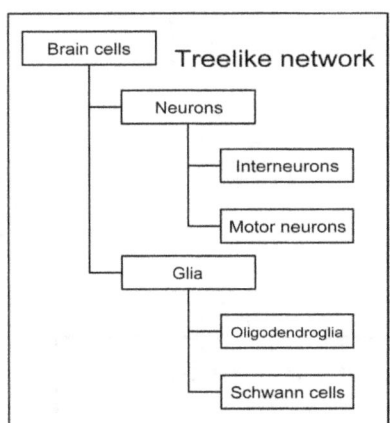

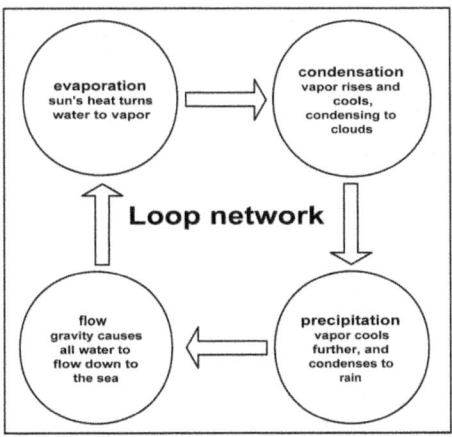

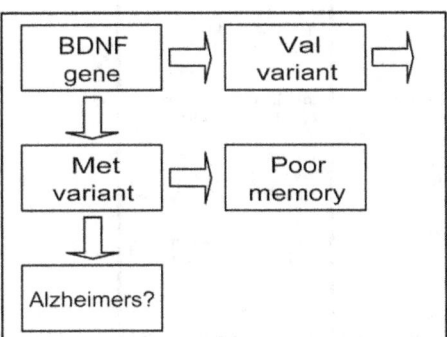

Fishbone diagram

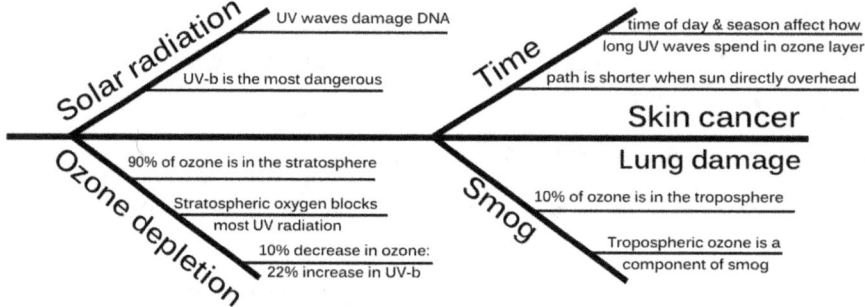

Venn diagram

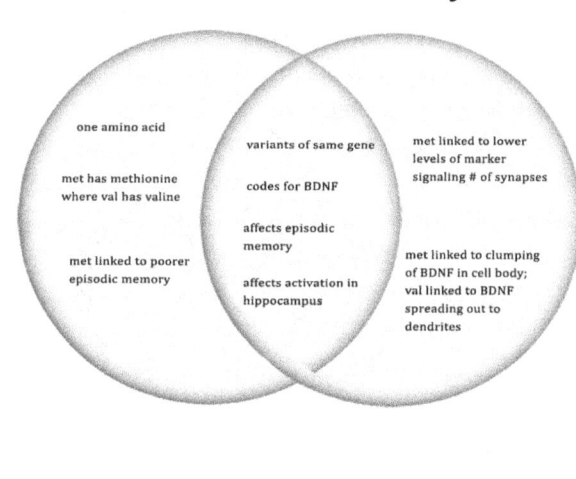

T-chart

Gene for Memory

Similarities	Differences
variants of same gene	one amino acid
codes for BDNF	met has methionine where val has valine
affects episodic memory	met linked to poorer episodic memory
affects activation in hippocampus	met linked to lower levels of marker signaling # of synapses
	met linked to clumping of BDNF in cell body; val linked to BDNF spreading out to dendrites

Spider map

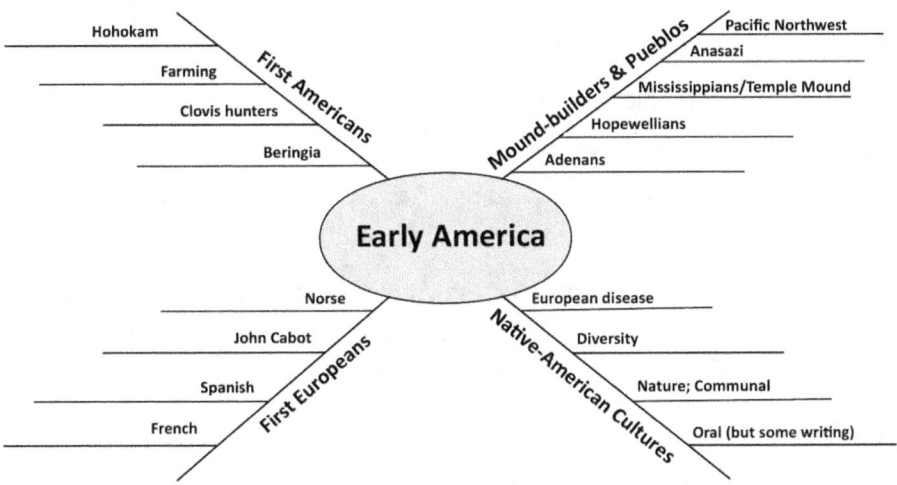

Refutational map

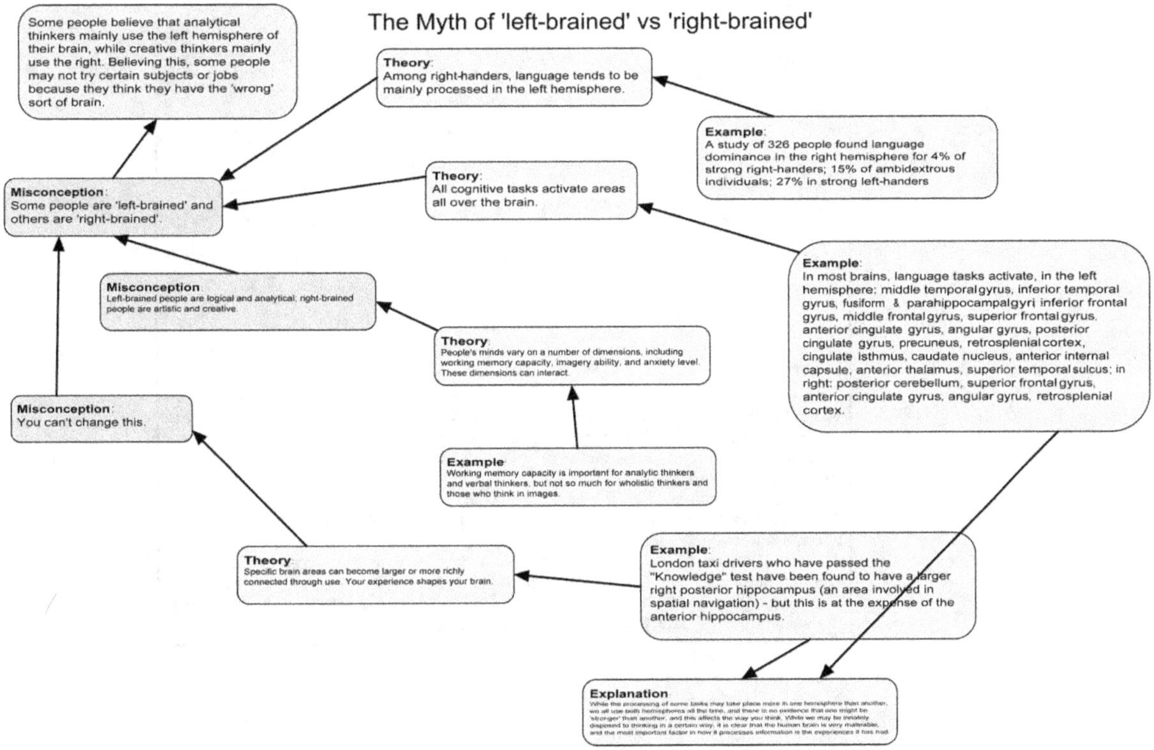

As you can see, there are many formats of graphic organizer you can use to summarize (and this is not a complete list). What matters is finding the one that is the best fit, both for the information and for you.

Constructing multimedia summaries

Multimedia summaries aren't just a type of illustration that you might see in a textbook or on the Web. Multimedia summaries are something that you can create yourself, as a means of taking notes and helping your understanding.

The key thing to remember in constructing your own multimedia summaries is that words and images must be tightly integrated, with very concise text, and everything labeled. Compare the two examples below.

What not to do:

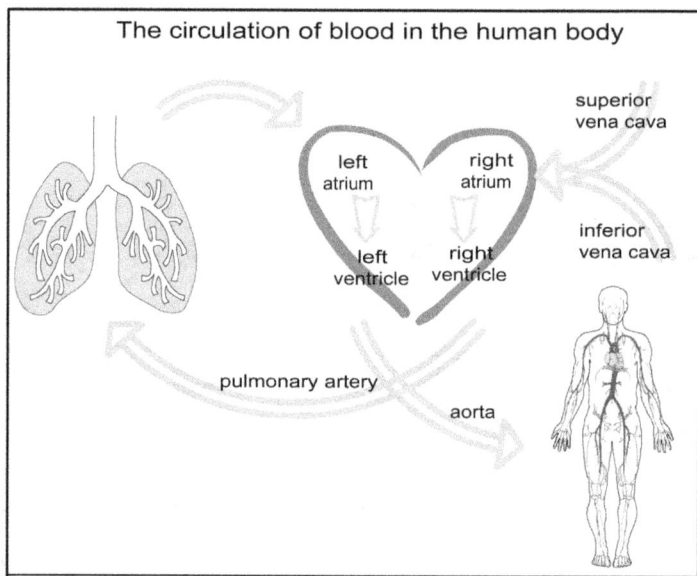

The problem with this example is that everything is happening at once, and the text is lumped all together, separate from the illustrations. We need to break it down into steps, with the appropriate bits of text right next to the matching illustration.

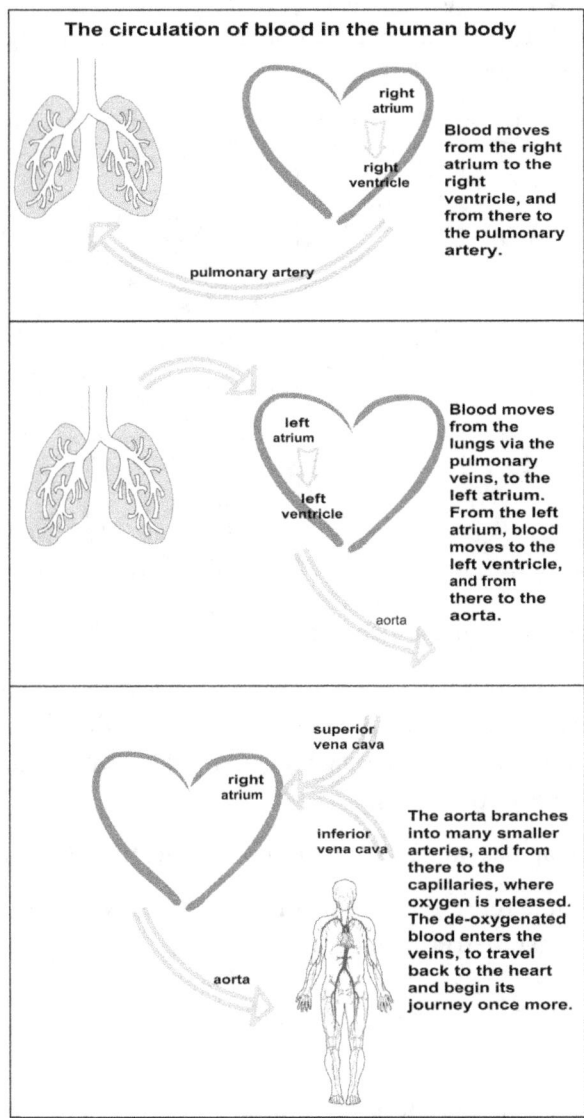

Remember: multimedia summaries are most useful when you're describing *process* (such as the water cycle, or the movement of blood around the body) rather than a set of facts.

Creating maps

If your text contains geographic information, as the early American text did, this is often best displayed in a map. The territories of the early American cultures, for example, didn't follow present-day state boundaries, making it difficult to simply attach a one-word label (such as 'Ohio'). It's best, therefore, to get a picture of the territory information (do note that I make no pledges about the accuracy of these maps!):

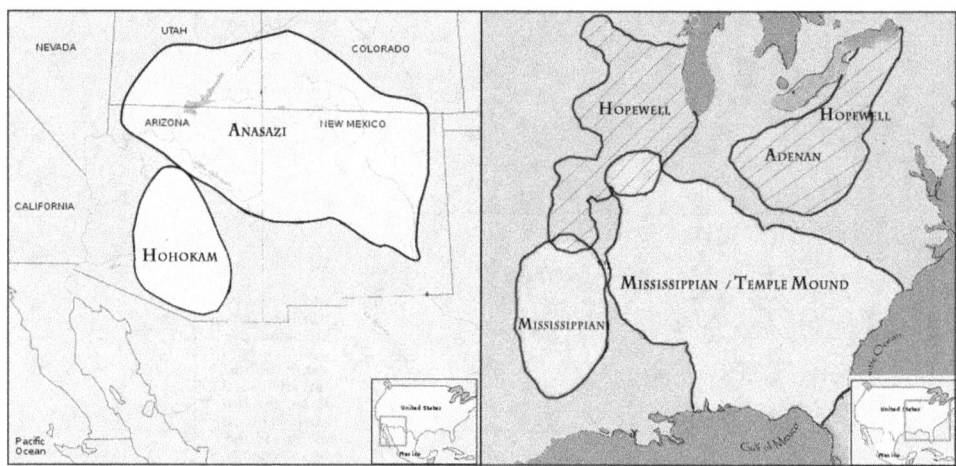

Remember, if you're drawing a map, that:

> - the important thing is the relationships *between* features, rather than the specific features themselves (e.g., that the Adenans are within the territory later claimed by the Hopewell cultures; the vicinity of the cultures to the Lakes)
> - a useful map has only a limited number of features (12-16)
> - icons (simple representative pictures) are usually better than arbitrary geometric symbols.

Main Points

Notetaking is a skill, and like all skills, it improves with appropriate practice.

The more difficult the notetaking situation (difficult text; a disorganized and fast-talking lecturer), the more you need good note-taking skills and good subject knowledge.

- Good subject knowledge helps you know what to select.
- Good note-taking skills help you format your notes appropriately.

Highlighting text as a means of taking notes is only useful when the text is very simple (for you), or you only want very specific information from it.

Note-taking is basically about summarizing. But summaries don't have to be simple lists:

- Choose a format that's compatible with the text structure.
- Where text is complex and won't fit a single structure, feel free to use more than one (e.g., the American history text as a table, map, and spider map).

Good notes are re-worked. Don't expect to achieve them on a single pass.

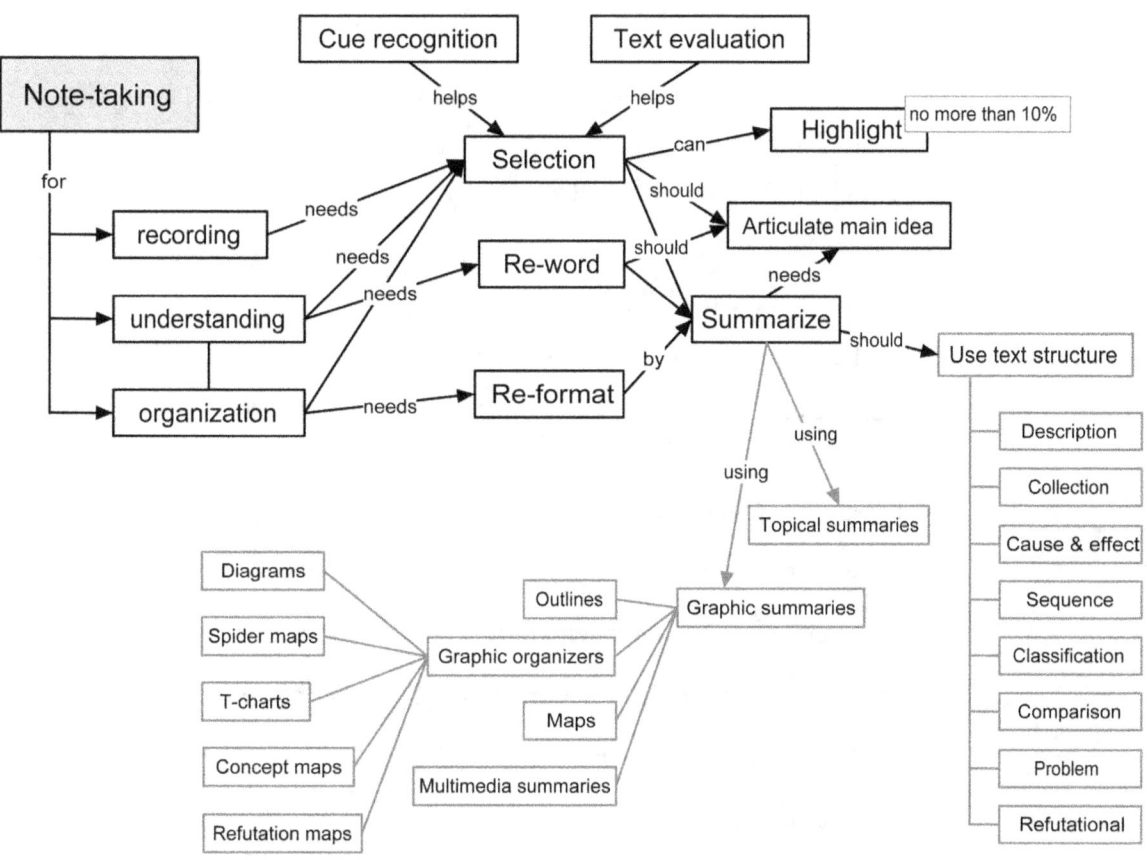

Taking Notes 121

Review Questions

1. Taking good notes is about:

    a. Writing down as much as you can

    b. Paraphrasing, organizing and making sense of the information

    c. Writing in a strict format

    d. Writing clearly

2. Selecting the most important information is more difficult when:

    a. There's a lot of information

    b. The lecturer talks with an accent

    c. The lecturer speaks quickly

    d. The material is poorly organized

    e. There's a lot of important information crowded together

3. Highlighting is only useful when:

    a. You're a novice

    b. You have a low WMC

    c. Less than 10% of the text is important

    d. The text is simple

4. A good summary:

    a. Contains only the most important information

    b. Contains all the topic sentences

    c. Contains the best phrases and sentences from the text

    d. Is written in your own words

    e. Is as long as it needs to be

5. To create a good summary, you should:

   a. Take notes as you read

   b. Read the text before taking any notes

   c. Draft it without looking at the text

   d. Always use the text structure as the framework of your notes

   e. Begin by writing down what you think is the main idea, in your own words

6. The best text structure is:

   a. Cause-&-effect

   b. Sequence

   c. Description

   d. Refutational

   e. None of the above

7. Match the graphic organizer with the text structure it is best associated with.

Description
Comparison
Cause-&-effect
Refutational

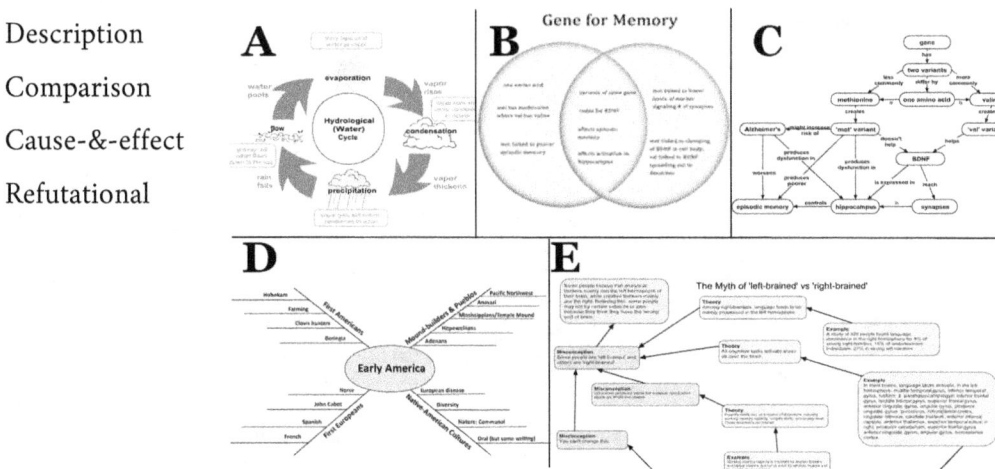

8. The text structure that is most effective for learning and recall is:

   a. Collection

   b. Problem-&-solution

   c. Comparison

   d. Refutational

   e. None of the above

Learning through Understanding

Making a good memory code is about selecting the right information, and connecting it to other codes in the network. A good memory code is richly connected. An expert is a person who has a richly connected network in a specific topic.

This isn't simply a matter of having a string of facts at hand. This is about understanding. We understand when we make connections. The more connected our network, the greater our understanding.

Understanding is rooted in the connections you make

Building your knowledge in a topic (as opposed to accumulating a set of isolated facts) is about making connections.

One of the best ways of making connections is by asking questions.

Asking questions

Here are some facts:

> Arteries are thick and elastic and carry blood from the heart. With the exception of the pulmonary artery, that blood is rich in oxygen.
>
> Veins are thinner, less elastic, and carry blood to the heart. With the exception of the pulmonary veins, that blood is rich in carbon dioxide.

We can easily turn these facts into questions:

Why are arteries elastic?

Why are arteries thick?

Why do arteries carry blood away from the heart?

Why do arteries carry blood that is rich in oxygen?

Why are veins less elastic?

Why are veins less thick?

Why do veins carry blood to the heart?

Why do veins carry blood that is rich in carbon dioxide?

But better questions will help us make the connections between the facts explicit:

1. **Why** do arteries need to be thicker and less elastic than veins?

2. **Why** do arteries carry blood away from the heart?

3. **Why** do arteries usually carry the blood that is rich in oxygen?

4. **Why** should blood be rich in oxygen?

5. **Why** is blood sometimes rich in oxygen and sometimes rich in carbon dioxide?

6. **Why** do the pulmonary artery and veins reverse the normal rules?

Do you see how these questions relate the facts to each other?

The 'facts' as written don't explicitly give us the answers to these questions, so the first thing to do is see if you can use your general knowledge to think of some plausible answers.

What is the one thing everyone knows about the heart? That it's a pump. It pumps your blood.

How do pumps work? In bursts, rather than as a smooth continuous flow.

This means blood must come *from* the heart in spurts, meaning that the flow will vary in volume and speed. That's why the vessels carrying the blood from the heart need to be stronger and more elastic.

What about the connection between blood and oxygen? Here's another piece of general knowledge you may know: blood is how oxygen reaches our tissues. You may also know the related fact, or can surmise it from this: blood is also the means by which the waste product carbon dioxide is removed from our tissues.

These facts explain why blood in the arteries (carrying blood *from* the heart) is oxygen-rich, while blood going *to* the heart (in the veins) is rich in carbon dioxide.

This also explains why the pulmonary artery and veins reverse the usual rules. The pulmonary artery carries the blood from the heart to the lungs, and the pulmonary veins carry the blood from the lungs to the heart. Blood coming from the heart must be carried by arteries, because of the pumping action. Therefore one particular artery — the one sending the blood from heart to lungs — must carry "dirty" blood. And because blood flowing into the heart must be carried by veins, some veins — those bringing the "clean" blood back to the heart — must carry clean, oxygen-rich blood.

Good questions:

> - direct your attention to the important details
> - require you to *integrate* the details in the text.

There are two main questioning strategies for making connections:

Elaborative interrogation involves taking a fact, asking yourself why it's true, and attempting to answer your question. This is what I did in the above example.

To be effective:

> - you need to attend to the information in your memory bank that is

consistent with the information to be learned — not the stuff that may be inconsistent.

> asking "Why is this true?" helps you focus on the consistent information.

> you need some relevant knowledge and understanding of the topic.

Self-explanation involves you explaining the meaning of information to yourself while you read. This is:

> a more flexible strategy than elaborative interrogation

> very easy to learn

> less reliant on prior knowledge than elaborative interrogation, so more suitable if you're new to a topic

> useful when the text is demanding.

To use this strategy, you simply ask yourself what the fact (question or paragraph) means to you. What new information does it provide? How does it relate to what you already know? Does it help you understand something better? Does it raise questions in your mind?

Here's the first paragraph of the blood flow text again:

> We all know that blood flows through our body in continuous motion, and that our heart is the pump that drives this motion. But the circulatory system is best understood not as a single system but in terms of its three constituent parts — pulmonary circulation (involving the lungs), coronary circulation (involving the heart), and systemic circulation (involving the blood vessels).

Here's how you might question yourself after reading that:

> Did I know that blood flows through our body in continuous motion? I guess so, though I never thought about it before. I knew the heart beats and that my heartbeats and the pulse in my wrist are connected, so I guess I knew the heart was driving the blood around my body in bursts. I knew circulation involved both the heart and blood vessels, but I didn't know the names for these parts (so I'll highlight those), and I hadn't thought of it including the lungs. Why does blood circulation involve the lungs? It must have to do with air. I know that we need oxygen to breathe, and there's something I remember vaguely about oxygenated blood being redder, and carbon dioxide poisoning making blood darker. So I guess when we breathe in, the oxygen somehow gets from the lungs into the blood.

Making comparisons

Another way of making connections is through comparisons.

To make connections through comparison:

- look for similarities
- look for differences
- seek analogies.

Differences can be

- related to a common structure (e.g., a dog has four legs but a human two)
- unrelated (e.g., a dog has four legs but humans can talk).

So, for example, veins and arteries have the following similarities:

- both carry blood
- both are hollow tubes
- both connect to the heart
- both connect to the lungs.

and the following differences:

- veins are thinner than arteries
- veins are less elastic than arteries
- veins carry blood to the heart, while arteries carry blood from the heart
- veins carry blood from the lungs, while arteries carry blood to the lungs
- most veins carry de-oxygenated blood, while most arteries carry oxygenated blood.

Spelling these out to yourself will help you make the right connections, and remember the information.

Analogies

One familiar type of comparison is the **analogy**. Analogies are about using an example you know well to help you understand something you don't understand very well.

Analogies are based on a similarity of structure, of relations between objects. Unfortunately, because we rarely file memories by such deep attributes — because we tend to think in terms of surface details — finding good analogies is hard.

Analogies can either:

> ➢ compare an unfamiliar instance to a familiar one, or
>
> ➢ compare two partly understood situations.

The second comparison is usually easier for the novice.

To do this:

1. view two examples side-by-side
2. assign specific correspondences between the elements of each example (**correspondence list**)
3. describe them as a unit, by comparing one to the other (**joint interpretation**).

For example:

1. The "BDNF" gene has two variants that differ in one amino acid, creating genes that differ in their effect on BDNF distribution to the synapses, thus affecting memory
2. "Principal" / "principle" has two variants that differ in one letter, creating words that differ in meaning

Correspondence list:

Situation 1	Situation 2
gene	word
amino acid	letter
effect	meaning

Joint interpretation: The gene (word) has two variants that differ in one amino acid (letter), thus affecting behavior (meaning).

To benefit from analogies, you need to work them out from specific examples.

Concept maps

One note-taking format is especially good at helping you make connections: concept maps.

A concept map is a diagram in which labeled nodes represent concepts, and lines connecting them show the relationships between concepts. Here's a concept map for the brain cells text:

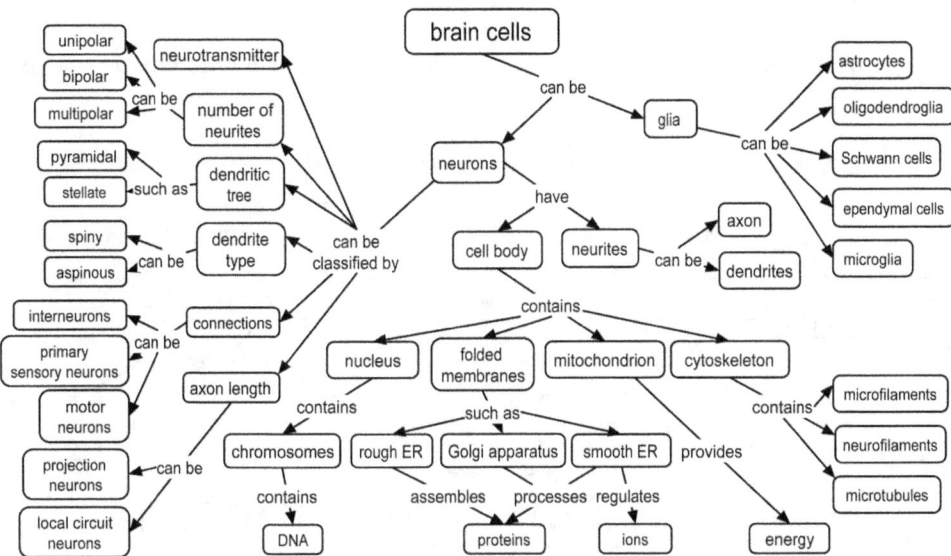

Note that a "concept" can be just about anything: a noun, a verb, an adjective, an adverb, a descriptive phrase.

When concept maps are useful:

> at the very beginning of your study, to prime your mind for what you are about to learn

> in the middle, to take notes and improve your understanding (make connections)

> at the end, to review what you have learned.

Sketching a concept for priming or review is quite straightforward.

For example, say you were about to read (or listen to a lecture) about Vivaldi. You could have sketched a map outlining what you already knew about him beforehand, like this one on the next page.

Learning through Understanding

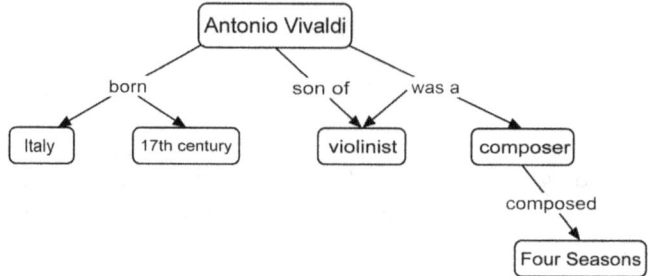

Later on, as a means of review, you might produce a more detailed map, such as this:

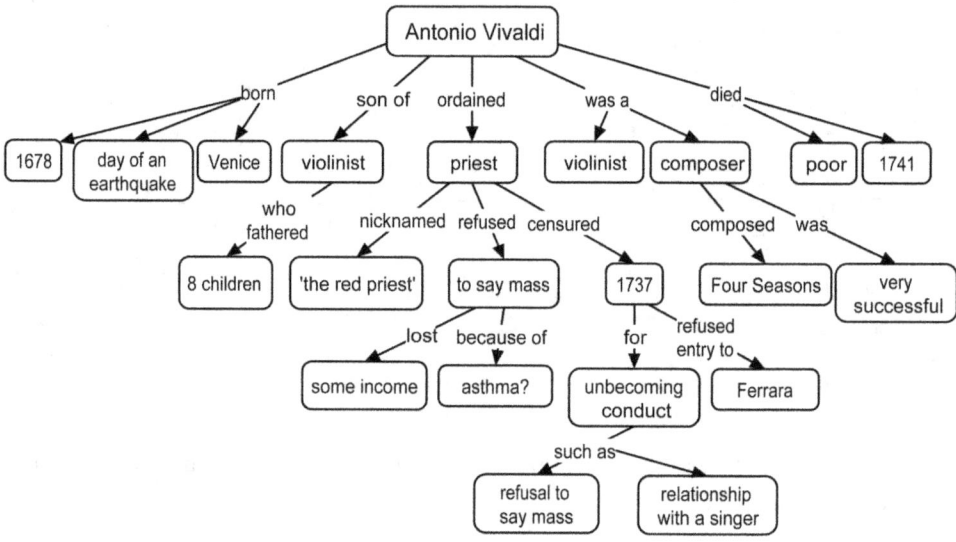

As a means of notetaking, concept maps are most useful for organizing your knowledge. They help you sort out the big picture; they show you where everything fits. They're not so good for detail.

Where concept maps excel is in showing you what you don't properly understand. They are therefore particularly recommended for complex topics.

Creating a good concept map is something that has to be worked for; it takes time, and repeated efforts.

On the next page, for example, you can see an early concept map I drew for the difficult 'memory gene' text.

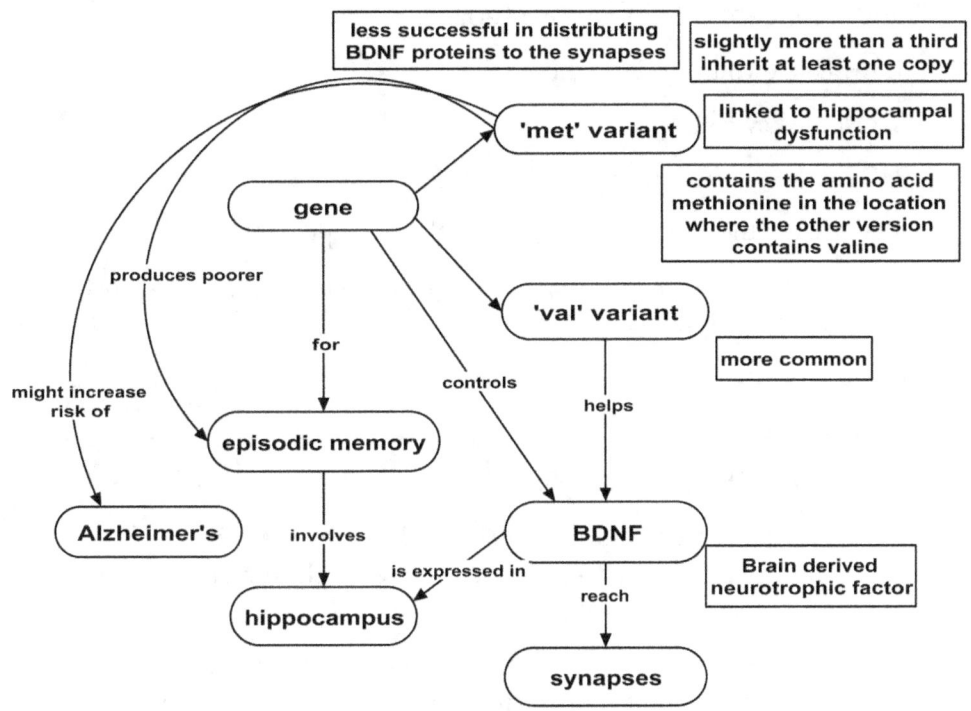

And here is my final version (*Effective Notetaking* covers the process in detail).

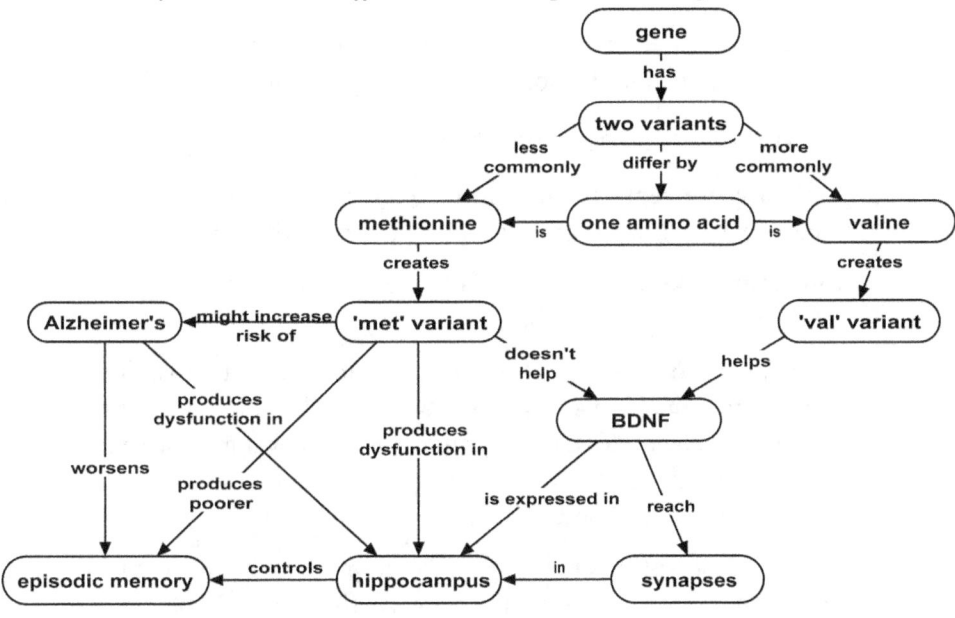

Learning through Understanding

Building a concept map, step-by-step:

1. Articulate your focal question
2. List the key concepts
3. Describe the attributes of these concepts
4. Articulate the relationships between the concepts
5. Order the concepts in a rough hierarchy from most general to most specific, in this context
6. Draw your first map
7. Evaluate it:

   ★ Are all the links clearly labeled?

   ★ Are concepts as short as possible? (short phrases are fine; if they're full sentences, it's a sign you're overloading the concept — it needs to be broken up)

   ★ Are there good cross-links (ones from one section of the map to another), but not so many it's confusing? (remember, almost everything can be seen to be connected; you have to be selective)

   ★ Does the map cover the material?

   ★ Does it answer the focal question? (novices are prone to wandering from the point; a good map is a focused map)

Common problems and their solutions:

➢ **Too many concepts**: break your map down into smaller maps, with one 'overview' map that shows how each map fits together.

➢ **Too many links**: again, you need to create separate maps for some of the subtopics.

➢ **Descriptions on the links too long**: probably a sign that you're loading too much on a single concept — think about whether the concept can be divided into smaller concepts; or it may be that you don't clearly understand the relationship between the concepts — think about it a little more.

➢ **Concept descriptions too wordy**: again, you map be loading too much on the concept — think about whether it should be divided into smaller concepts.

- **Too many maps**: there is a happy medium! If some of your smaller maps are too small (only contain 3 or 4 concepts), think about merging them with another.

- **Not enough detail**: consider adding more concepts.

- **Doesn't cover the material**: probably a sign that you haven't clearly and accurately articulated what the concept map is about — try and formulate the question you want the concept map to deal with; if you've got that right, that question should include as its subject the concept that will be the starting point for your map. If that doesn't work out, reformulate your question.

You may want to add specific examples to help you clarify a concept. If you do that, don't enclose the examples in ovals or boxes, because you don't want them mistaken for concepts.

Mind maps

Mind maps are a specialized form of concept map popularized by Tony Buzan. Here's a mind map version of the brain cells text. Note the ways in which it differs from the concept map of the same information.

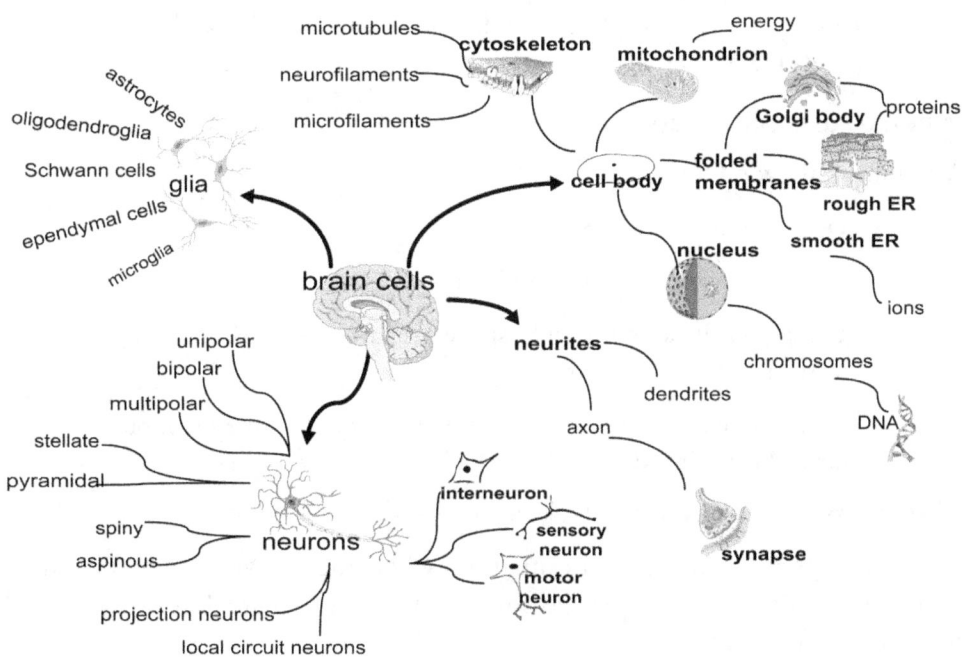

How mind maps differ from concept maps

A mind map has four essential characteristics:

> ➤ The subject is crystallized in a central image
>
> ➤ Main themes radiate from it as branches
>
> ➤ The branches comprise a key image or key word
>
> ➤ The branches form a connected nodal structure.

Keywords — what Buzan calls BOIs (Basic Ordering Ideas) — are the 'concepts'.

Important differences between mind maps and concept maps:

> ➤ In a mind map the main themes are connected only to this single central image — not to each other.
>
> ➤ In a concept map, there are no restrictions on the links between concepts.
>
> ➤ The connections between concepts in a concept map are labeled — they have meaning; they're a particular kind of connection.
>
> ➤ In a mind map, connections are simply links; they could mean anything.
>
> ➤ Mind maps are more colorful and pictorial.
>
> ➤ Concepts in a concept map, on the other hand, can be (and usually are) entirely verbal.

When to use mind maps

Mind maps are good for:

> ➤ generating ideas
>
> ➤ helping you sort out the main ideas
>
> ➤ getting your head in the right space preparatory to listening to a lecture or reading a text
>
> ➤ doing a quick review — checking that you have all the main points down before a test.

They're less useful as a means of taking notes.

Concept maps are better suited to situations where the concept is to be shared with others. Mind maps are very personal, and harder to share (less easily understood by others).

Problems people often have

- Staying 'on track' (being de-focused is fine when 'brainstorming', but not when summarizing): your Central Image is probably not properly focused, or you haven't got your BOIs quite right.
- Coming up with your BOIs: if you're trying to summarize a textbook, or a significant portion of one, then chapter titles are a good place to start, with sub-headings being used for branches.
- Making sure you've covered everything: check it against a more structured outline.
- Too many connections between nodes: you're probably better to go with a concept map.

Don't get bogged down in the "rules" of concept maps and mind maps. The essential characteristic is that you are drawing a spatial representation (our brains are good at remembering spatial information) that connects concepts in a useful way. What works best depends on both you and your material.

What this means for notetaking

Effective notetaking not only involves selecting the important information, but also involves making it meaningful to you.

Making information meaningful is about connecting new information to existing knowledge:

- The more connections you make, the better you will understand the information.
- The more connections you have, the more entry points you have to the information, therefore the easier it will be to find in your memory.

For example, from those *why* questions we asked about blood flow, we can build a **multi-connected cluster** (see next page). Notice the presence of three 'general knowledge' bits of information:

- that our heart is a pump / that our blood 'pulses'
- that we breathe in oxygen
- that we breathe out carbon dioxide.

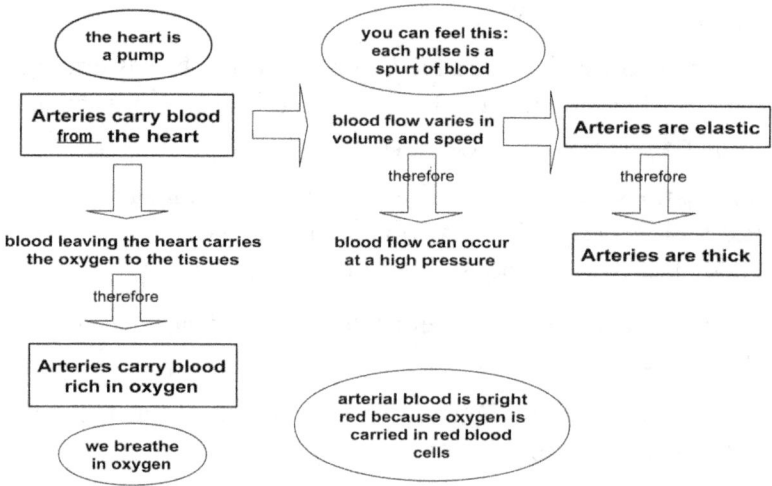

These are what connect the new information to information you already have in long-term memory.

Facts that you already know very well and have no trouble remembering act as **anchor points**.

Anchor points are all potential entry points to the cluster.

The more anchor points, the more entry points you have to the information, therefore the easier it will be to find in your memory.

The more anchor points, the more meaningful the new information becomes, and the more easily you will remember it.

This then, is how you can judge the value of a particular notetaking strategy in a particular situation — ask yourself:

> ➢ Does it help me connect the facts together?
>
> ➢ Does it help me connect the new information with information I already have?
>
> ➢ Does it make any connection with facts I already know very well, and am unlikely to forget?

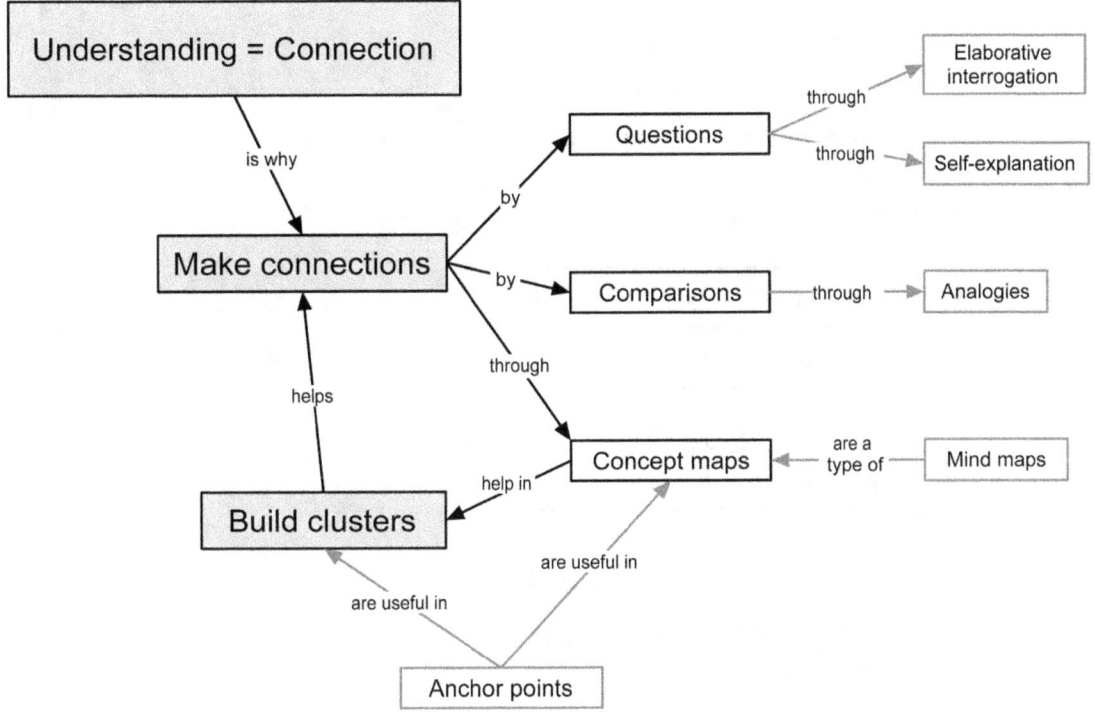

Review Questions

1. Asking questions helps you make connections. Good questions:

   a. Are short

   b. Ask why things are true

   c. Ask what things mean to you

   d. Point to the important details

   e. Make you put information together

   f. Require good background knowledge

2. Making comparisons helps you make connections between things that are:
   a. Similar in appearance
   b. Different in appearance
   c. Similar in structure
   d. Different in function
   e. Alike in some ways and different in others

3. Fill in the missing parts of this correspondence list.

Situation 1	Situation 2
Atom	
	Sun
Electrons	

4. Concept maps are most useful for:
   a. Brainstorming
   b. Priming
   c. Revision
   d. Making connections
   e. Taking notes in lectures

5. Mind maps are most useful for:
   a. Brainstorming
   b. Taking notes
   c. Revision
   d. Sharing with others

Reading on the Web

One of the complications of study in this modern age is that we are at a time of transition, with an increasing amount of our study material presented to us on the Web or in the context of some other non-linear environment, while we are only beginning to grasp how that impacts learning. This affects both ends of learning: those creating the study materials are as yet unversed in how to present them in an effective way, and students have little training in how to study such material.

Information in books and lectures is presented in a 'linear' fashion: each piece of information presented one after another in a specific order.

Computers have opened up another possibility, that of non-linear environments. The internet, where you move via hyperlink, where your path can be in any direction, is the pre-eminent example of a non-linear environment.

One of the big problems with non-linear environments is that they place much greater demands on working memory. Non-linear environments also tend to be multimedia, adding to the cognitive load.

Strategies for dealing with non-linear environments usually focus on reducing cognitive load.

While non-linear environments may sometimes need different skills than those used in more familiar linear environments, the main problem is that a non-linear environment puts higher demands on your general study skills.

In other words, you might be getting by with poor skills in less demanding situations, but your lack of good skills will be revealed in more demanding contexts.

The following general study skills are thought to be particularly useful if you want to learn effectively in a non-linear multimedia environment (also called a **hypermedia environment**):

- analyzing the learning situation
- setting meaningful learning goals
- evaluating your emerging understanding of the topic
- determining whether the learning strategy is effective for a given learning goal
- monitoring understanding and then appropriately modifying plans, goals, strategies, and effort in relation to contextual conditions
- reflecting on the learning episode and modifying existing understanding of the topic.

What all this means is that **self-monitoring** and **goal-setting** are of paramount importance in successfully navigating this type of environment.

Navigating a hypermedia environment

Navigation is the heart of the reason why these skills are so critical in a hypermedia environment. Here are some examples of e-learning websites to show why.

Example 1:

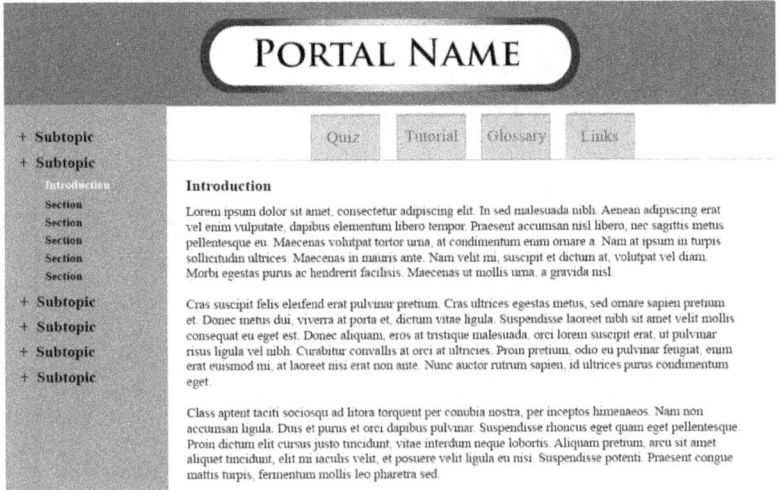

Instructions to the user typically run something like this:

> [This site] is divided into several topic areas, each with a tutorial and an assessment quiz. Students may start with the tutorial or the quiz, and may switch between them at any time. Students may use as much or as little of the [site] as they like.

This example is a contained site, with six subtopics, and from three to six sections within each subtopic. There is one quiz per subtopic, and 'Tutorial' simply refers to text — one page per section. With about 25 sections in total, there isn't a huge amount of material to get lost in, but because there's no means of knowing what you've done, with the possible exception of link color change (the links in the menu may change color after you've used them, but they probably won't, and even if they did, that doesn't tell you whether you simply had a quick look, or actually studied the material), your best strategy is probably to begin at the top of the menu list and work your way down, doing the quiz at the end of each subtopic.

In other words, the only clear way of getting through this material is to treat it in a linear fashion, as you would a book (but without the same easy ability to annotate or stick in casual bookmarks).

Example 2:

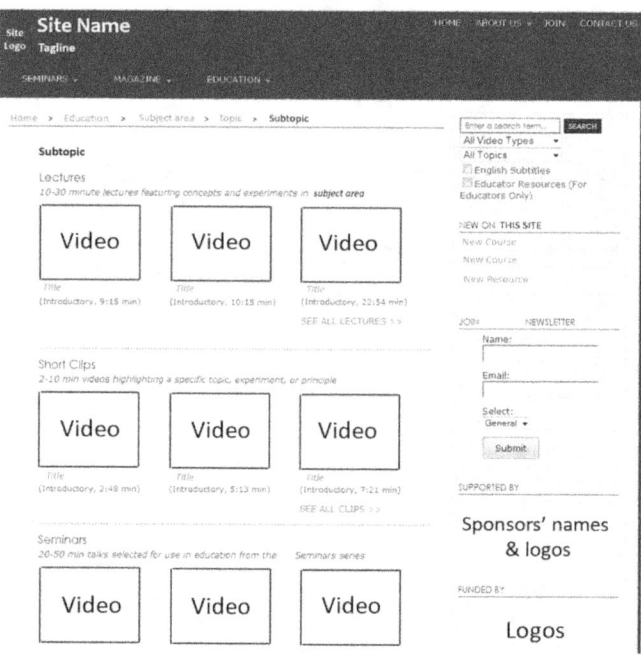

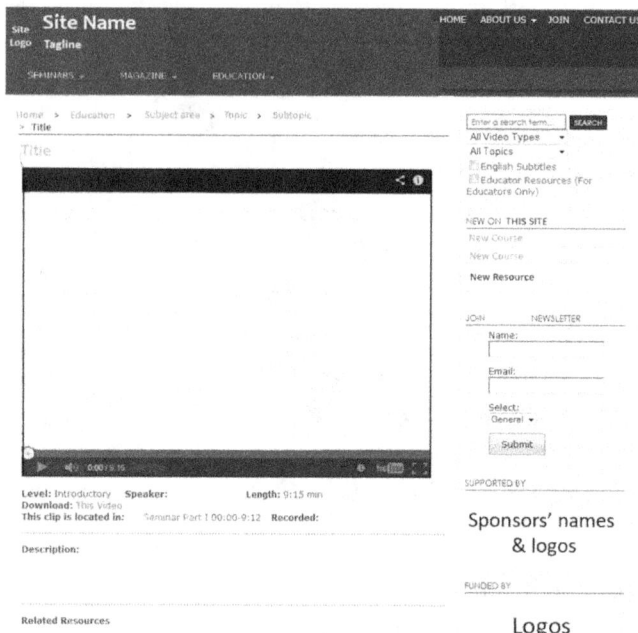

Clicking on one of these videos will get you a page like this on the left.

Here there's no attempt to support learning, other than by providing information. There's no navigational support, and no opportunity to assess your learning.

Example 3:

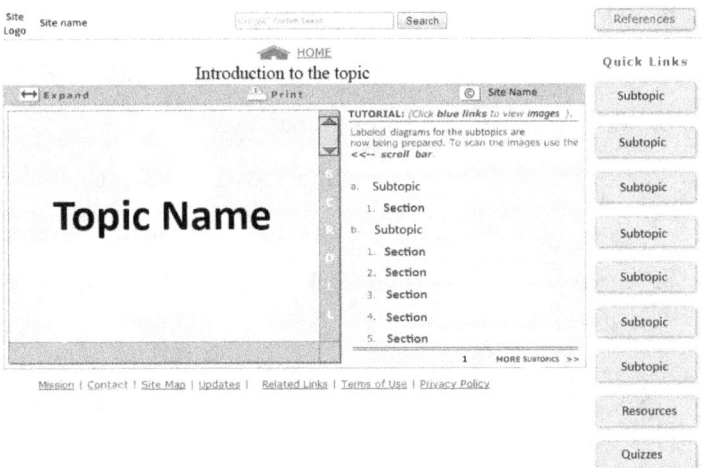

Here images again dominate, but this time the images are static. The large rectangle usually holds a diagram, and the "Tutorial" box usually contains a brief text. There are a lot of sections, but the "Quick Links" bar doesn't change — once you're within a subtopic, your only means of navigating within it is to click the "Previous" and "Next" buttons, or use the breadcrumb to return to the subtopic's index. There is no way of keeping track of what you've covered.

So what do you do, to help you keep track of what you've done? The obvious way is through your notes, by using the website structure. Note down the name of each section/page as you go, even the ones you don't take any notes from (make a note that it wasn't useful, or that you might want to return to it at some particular time).

As with textbooks, it's also helpful to get an overview first. Let's say that all these websites are on the same topic. With the first one, the navigation bar only shows the section titles for one subtopic at a time. Similarly, the other websites also only show part of their wares at a time. You may find it helpful to write down the various lists of subtopics/sections, so you can work out your plan of attack. This will also help you activate your relevant existing knowledge, and refine your questions.

These examples are all quite simple hypermedia environments: predominantly linear, predominantly a single type of representation. These are the type of situation you're most likely to come upon on the Web. The Web itself, however, and the more sophisticated hypermedia environments you might be presented with as part of an educational or training program, are less linear and more diverse.

If you are attempting to learn from diverse sources on the Web, you need to:

- ➢ set your goals at the start — spell out the questions you need to answer
- ➢ modify or add to these questions as you go
- ➢ keep track of what you've done / where you've been
  - ★ open new tabs for new sites, especially when you side-track
  - ★ create a folder in your bookmarks for the topic and collect all the useful sites there
- ➢ try and get a 'big picture' view of the material before delving into the detail.

Problems & benefits of animations

In the second example website, all the material was in the form of video clips, which are essentially lectures with slides. Lecture videos certainly have important advantages over actual lectures:

- ➢ You can replay them as often as you like.
- ➢ You can pause them.
- ➢ You can rewind them.

However, they also have major disadvantages over text formats:

- ➢ You can't skim them.
- ➢ You can't easily go back and forth between related bits of information.
- ➢ You can't easily compare images that aren't in the same frame.

Videos, then, need to be used judiciously, not as the sole format.

Animation is often part of hypermedia environments, but note that, although animations are usually more fun than static diagrams, they have several disadvantages:

- ➢ You can't control the pace.
- ➢ You can't as easily focus on one thing at a time.
- ➢ You're more likely to passively watch rather than truly engaging with it.

Feel free to enjoy animations, but pay careful attention to how well an animation has helped you understand; it may be necessary to seek out other sources to help you understand particular components of the process demonstrated, before returning to the animation for the 'big picture' understanding.

How to learn effectively in hypermedia environments

The two main difficulties in dealing with hypermedia environments lie in:

- the need to coordinate multiple representations of information (images, text, audio, and so on)
- the need to determine the right instructional sequence from the several possible.

Hypermedia environments that are part of an educational or training program should (if they're any good) provide most of the following tools:

- advance organizers
- navigation maps
- search tools
- questions at appropriate places:
    - ★ to monitor understanding
    - ★ to activate any knowledge you already have
    - ★ to help you integrate different pieces of information
    - ★ to help you remember what you're learning
- prompts to use appropriate strategies
- explicit goals at appropriate points, to keep you motivated and help you monitor your progress
- a planning component to help you plan your learning activities.

You should take full advantage of these supports if available!

If your hypermedia environment doesn't include these (and you don't have the option of finding a better one), you should provide as many as these for yourself.

What you can do:

- Make up questions and try to answer them.
- Personalize the information — make it about "I" and "you" (e.g.,

"your heart" rather than "the heart"). (Research has shown that such personalization makes it easier to understand and remember.)

> Try and see all parts of a problem before tackling it.
> Try and get feedback after every problem.
> Bring your focus down to what you can manage
>   ★ ignore less-relevant information
>   ★ chunk as you go.
> Try and connect your new chunks to information you already know.
> Write down your goals and check back as you go, to make sure you're still on target.
> If you are having a hard time understanding the information provided, seek out other sources to build your knowledge of the topic, before returning.

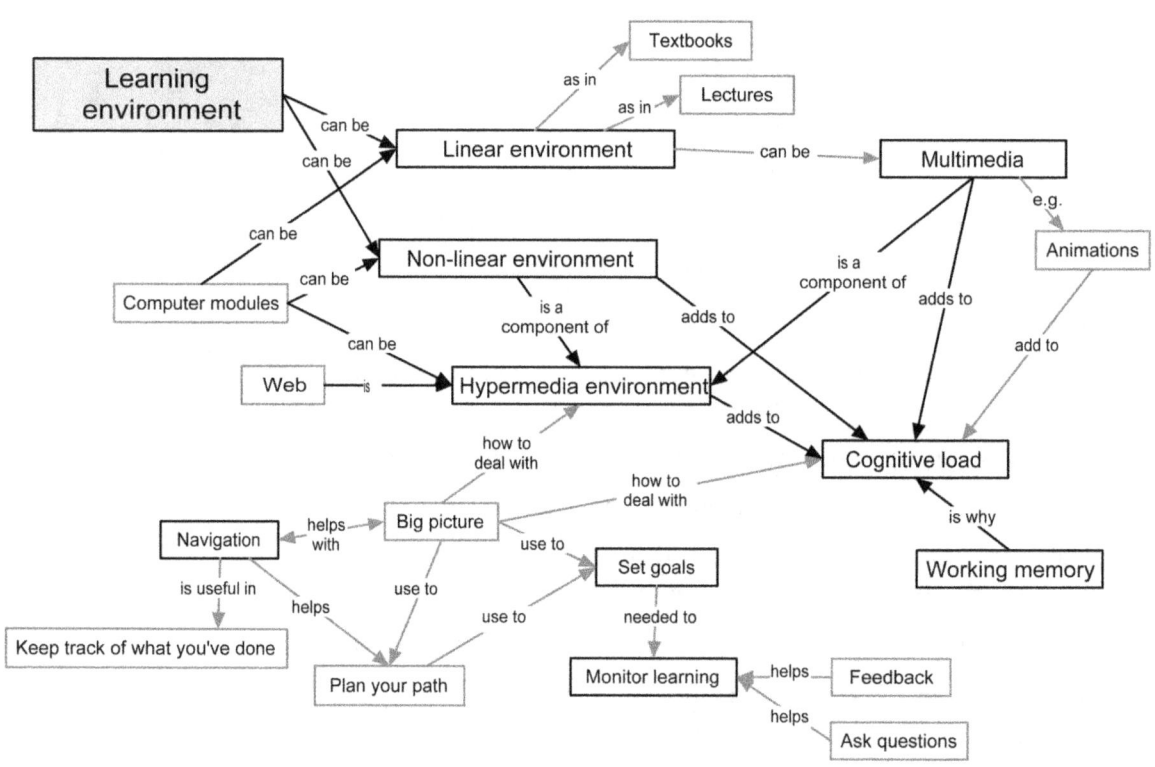

Review Questions

1. Hypermedia environments:

    a. Are fast-moving

    b. Involve combinations of pictures and text and sound

    c. Put high demands on working memory

    d. Follow a linear path

    e. Are navigated by hyperlinks

2. Effective study in hypermedia environments requires:

    a. Good self-monitoring skills

    b. A high working memory capacity

    c. Good goal-setting

    d. Use of strategies that reduce cognitive load

3. The two main difficulties with non-linear environments are:

    a. Navigation

    b. Lack of feedback

    c. Animations

    d. Cognitive load

4. Videos and animations have several advantages:

    a. You can scan them quickly

    b. You can replay them as often as you like

    c. You can easily compare related slides / images

    d. You can control their pace

    e. You can pause and rewind them

5. Videos and animations have several disadvantages:

   a. You can't scan them

   b. You're more likely to watch them like a TV, rather than engaging with the content

   c. You can't easily compare related slides / images

   d. You can't control their pace

   e. Because so much is going on at the same time, it's harder to focus on one thing

Getting the Most Out of Lectures

Taking notes of something you're listening to, whether it be a lecture or a video or a podcast, is very different from taking notes from a text.

Taking notes from audio is a two-stage process, for the recording element is now much more important, but the need to transform the notes remains.

The first stage, the recording, is one that overwhelms many students. What you need to understand is that this recording is a cognitive skill, and like all skills is very much dependent on practice. After you've read the later section on Skills, come back to this chapter and think about it with that in mind.

How lecture notes are different from textbook notes

- You have no control over the rate of presentation. Notetaking for encoding (as opposed to recording) needs a rate of 100 words per minute or slower (normal rate of speech is around 150 words per minute).
- You have no ability to go back over information you don't understand or didn't catch (apart from asking the presenter to repeat himself or explain more fully — something many students are shy of doing).

Lecture notes are therefore important more for *recording* than *encoding* — which means you have to work on them later to get value out of them.

Surveys have found:

- the average student records an average of only 60% of the information they should record
- this varies a lot by subject, and by lecturer, and can be as low as 20% of important information!
- students do best with information written on the board
- the most successful students record more of the information *not* written on the board
- students greatly overestimate the adequacy of their notes
- students tend to record less and less information as the lecture goes on.

Taking notes in a lecture is demanding in part because you have multiple things to attend to:

- your own hand-movements and the writing you're producing
- the words spoken by the lecturer
- the actions of the lecturer and any visuals displayed.
- also, perhaps, your phone and other electronic devices!

Are there special strategies for taking notes in lectures?

Concentrate on recording, don't try to make sense of the material unless and until you have:

➢ well-practiced notetaking skills,

➢ a reasonable degree of expertise in the subject,

➢ high working memory capacity, *and*

➢ the lecturer is well-organized and doesn't talk too fast.

If you have good typing skills, you can reduce cognitive load by typing rather than writing by hand.

You can also reduce cognitive load by blocking out other distractions (such as your emails or Facebook! But also unrelated thoughts, such as your evening or weekend plans).

Use as many abbreviations as you can — while taking notes from books, build up useful abbreviations for words that you use a lot, and especially for long technical words.

Here are some that I use:

#	—	number	envt	—	environment
beh	—	behavior	esp	—	especially
c	—	circa / about	exc	—	except
circs	—	circumstances	expt	—	experiment
cog	—	cognitive	fam[y]	—	familiarity
cogn	—	cognition	freq	—	frequent
dep	—	depending	freq[y]	—	frequency
devt	—	development	fship	—	friendship
devt[al]	—	developmental	gen	—	generally
diff	—	different	govt	—	government
disc[n]	—	discrimination	grp	—	group
divn	—	division	hr	—	hour

id	—	identity	reln —	relation
immed	—	immediate	relnship —	relationship
immed^ly	—	immediately	rem —	remember
impt	—	important	repr —	represents
incl	—	including	reprn —	representation
incr	—	increase	req —	requires
incr^ly	—	increasingly	resp —	response
indep^t	—	independent	rqmt —	requirement
indic	—	indicates	S —	subject
indiv	—	individual	sev —	several
info	—	information	shd —	should
ltd	—	limited	signif —	significant
mem	—	memory	sim^y —	similarity
motiv^n	—	motivation	situ —	situation
nec	—	necessary	stim —	stimulus
nec^ly	—	necessarily	stim^n —	stimulation
obj	—	object	suff —	sufficient
orgn	—	organization	sugg —	suggests
popn	—	population	thru —	through
poss	—	possible	trg —	training
prev	—	previous	usu —	usual
pt	—	point	yr —	year or your (depending on context)
Q	—	question	v —	very
recogn	—	recognition	wmc —	working memory capacity
ref	—	reference		

Successful notetaking in lectures requires you to be as selective as possible. Therefore it's vital that you prepare for the lecture.

Any outlines or handouts provided by the lecturer will help reduce the demands on your working memory — use them as a base for your notes.

Thoughtful selection is the key to good notetaking. Don't record information just because others are, or the lecturer seems to expect it.

Don't ignore important information that comes up during discussions.

Do follow up the lecture by paraphrasing and organizing your notes.

If the lecturer provides an online video or audio of the lecture, it's worth playing this *after* you've worked up your initial notes, and have a better idea of what to listen for.

If you have only the one presentation of the lecture, focus on the 'top-level' ideas.

Different approaches to lecture notetaking

Ideas students have about lectures:

> ➤ You should record everything.
>
> ➤ You have to 'crack the code' — work out what the lecturer's mysterious signals mean, that tell what bits of information are important.
>
> ➤ You should just try and soak up everything.
>
> ➤ Lectures are just one source of information.
>
> ➤ Lectures are a guide to all the material you have to learn.

Students who believe you have to record everything behave differently than those who are seeking the secret code; students who think everything they need is in the lectures behave differently than those who see lectures as just one source of information. Think about what you believe, because your beliefs affect the strategies you use, both in class and out.

Now think about whether those beliefs are the best ones you can have. Think, too, about whether your beliefs should always be the same.

Effective notetaking requires flexibility — you need to adapt your strategies to the subject and the lecturer.

The lecturer's level and style of organization are particularly important.

- ➤ A topical outline or matrix may be best for more formally organized lecturers.
- ➤ A loose concept map or mind map may be best for more disorganized or informal lectures.

Especially when the lecturer is disorganized or gives too much information too quickly, don't disdain collaboration with other students to help you all put together a more complete set of notes.

Main Points

Unlike taking notes from a textbook, the main value in lecture notetaking is to *record* the important information.

Students record on average only 60% of the important information.

Students are less likely to record important information:

- ➤ later in the lecture
- ➤ if it's not written on the board
- ➤ if it's not expressed clearly
- ➤ if it's covered too quickly.

Success in lecture note-taking rests on reducing cognitive load.

You can reduce cognitive load by:

- ➤ reducing distraction
- ➤ practicing your writing/typing skills
- ➤ practicing your note-taking skills
- ➤ preparing for the lecture, by:
    - ★ building up your expertise in the topic (if only by doing the background reading!)
    - ★ thinking about: what is going to be discussed
    - ★ what topics/aspects will be most important
    - ★ how the lecturer typically organizes the lecture
    - ★ how you might best format your notes.

Remember that recording is only the first step. To benefit from your notes, you need to then turn them into good notes, by paraphrasing, re-formatting, and re-organizing them.

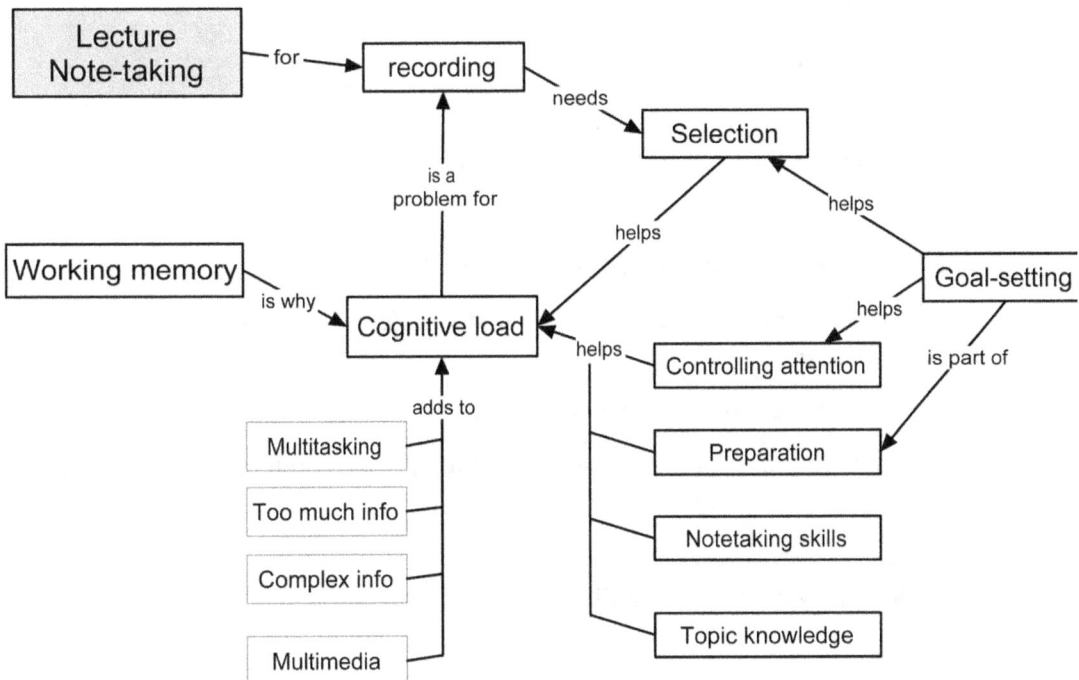

Review Questions

1. Taking notes in lectures is demanding because:

    a. Lecturers always speak too fast

    b. You have no control over rate of presentation

    c. You have no ability to skim, or backtrack

    d. It is less obvious which details are important

    e. You have to concentrate on multiple things

2. The most important information will be written on the board

   a. Always

   b. Not necessarily

   c. Reliably

   d. Never

3. Choose what information to write down by:

   a. Watching what other students do

   b. Listening for the lecturer's instructions

   c. Relating the information to what you know and considering it in terms of your goals

   d. Don't choose! Try to get everything down

   e. Assuming that anything important will be written on the board or the lecturer's provided notes

4. Taking notes successfully in lectures requires:

   a. Practice

   b. Preparation

   c. The ability to write quickly

   d. A high working memory capacity

Memorizing Verbatim

Mnemonics such as the Roman Room method are what "memory champions" use to memorize such things as the first 200 digits in pi. They are also what many memory improvement courses push. Their usefulness is, however, limited, as their function is to help you remember words and numbers verbatim.

Remembering exact word-for-word details is only necessary in certain specific circumstances. Still, these circumstances do occur in study, and for this material mnemonics are invaluable.

Mnemonics

Our brains are designed to remember the gist, the essence, not every detail. When we forget details, we usually regard this as a failure. But it's not a failure, it's a feature. Extracting the important information is the heart of intelligence, but sometimes we need to remember details that are not the kind of information we find easy to remember: precise details that aren't particularly meaningful to us. In that situation, we have two main types of strategy open to us: **spaced practice**, and mnemonics.

Mnemonic strategies:

- are "artificial" memory aids such as stories, rhymes, **acronyms**, and more complex strategies involving verbal mediators or visual imagery
- do *not* help you understand your material — they don't help you make meaningful connections
- create arbitrary connections in order to make hard-to-remember information more memorable
- are a bridging strategy — something you use to help you collect and use information until your new network is big enough, strong enough, organized enough, to become understandable and memorable on its own
- are particularly useful in two situations:
    - ★ to remember new words (which are often largely arbitrary)
    - ★ to remember the order of items
- can provide retrieval cues for your anchor points.

Vocabulary

If you are learning new words, for example, some of them may meaningfully relate to words you already know, either in your native language or in another language. But many will not. That's because there is a large arbitrary component to any language. If there's no meaningful connection to be made, then the next best thing is to invent a connection. This is what mnemonic strategies are all about.

So, in this list of 10 German words:

- der Apfel
- der Spinat
- der Kürbis
- der Kopfsalat
- der Fisch
- die Banane
- die Karotte
- die Kartoffel
- die Zitrone
- die Tomate

some of these words are very obvious: Apfel/apple; Spinat/spinach; Fisch/fish; Banane/banana; Karotte/carrot; Tomate/tomato. There's no need to make a special effort to remember these. You can, with only a little effort, also make a meaningful connection between Zitrone and its meaning (lemon), once you replace the 'z' with 'c' — for in English we have citron, citrus fruit, citric acid.

'Kopfsalat' takes a little more work, but if you're familiar with the epithet 'dummkopf', and realize that 'Kopf' is head, then the meaning of 'lettuce' (salad head) is quite easy.

Meaningful connections for Kürbis (pumpkin), and Kartoffel (potato) are much less easily found. This is where an arbitrary connection, a mnemonic link, is useful. Thus, you might create a mental image of *potatoes* falling into a *cart* full of *offal* (beef kidneys, chicken liver, sheep's hearts, etc).

Or form a mental image of a round, orange *pumpkin* rolling along the *kerb*.

Order

The second use of mnemonic strategies comes when the information, whether meaningful or not, needs to be remembered verbatim in a specific order. Thus, we remember the order of the colors of the rainbow by the mnemonic ROYGBIV, or the order of mathematical operations by the mnemonic BEDMAS (or BODMAS or PEDMAS, depending on your nationality).

But order, a specific sequence, isn't restricted to the few processes or objects for which common mnemonics are in widespread use. When you want to remember a speech, or the main points for exam essays, or a sequence of historical or scientific events, order mnemonics can be invaluable.

Words vs images

- Mnemonics can be primarily visual or primarily verbal; most combine elements of both.
- Images are effective to the extent that they link information.
- Images are not inherently superior to words.
- Bizarre images are not necessarily better recalled than common ones.
- Effective images involve the elements interacting with each other.

Primarily verbal mnemonics include:

- acronyms & **acrostics (first-letter mnemonics)**
- rhymes & songs
- sentences that tell a very short 'story' (**story / sentence mnemonic**)
- ways to transform numbers into words (**coding mnemonic**).

Primarily visual mnemonics include:

- systems that involve mentally visualizing objects in a series of set places (**method of loci** / Roman Room method / journey method)
- systems that involve mentally visualizing objects interacting with other objects that represent ordered numbers (**pegword mnemonic**)
- mentally visualizing objects that are linked together in a chain (**link mnemonic**)
- creating an interactive image tying two representative objects together (**keyword mnemonic**).

Rhythm & rhyme

Some examples of mnemonic 'jingles' that help us remember common facts through rhythm and rhyme:

Thirty days hath September,
April, June and November.

In fourteen hundred and ninety-two,
Columbus sailed the ocean blue

Remember remember,
The 5th of November,
Gunpowder, treason and plot.

I before E, except after C

Rhythm and rhyme are helpful to the extent that they:
- create expectancies
- set constraints
- make repetition more pleasurable.

Similarly, simple songs can help you remember, if they:
- use a simple, well-known tune
- match words to the tune, with one note per syllable
- have predictable, meaningful sentences and phrases
- follow a predictable pattern and rhyme.

In other words, the mnemonic power of rhythm and rhyme lies in
- predictability, and
- the increased likelihood that you'll keep repeating it.

The best jingles are short and simple, with a strong beat.

Keyword method

The most widely useful mnemonic strategy is the keyword method.

The essence of this technique lies in the choosing of an intermediary word that binds what you need to remember to something you already know well.

So, for example, to remember that the Spanish word *carta* means letter (the sort you post), you select an English word that sounds as close to *carta* as you can get, and you make a mental picture that links that word to the English meaning — thus, a letter in a cart.

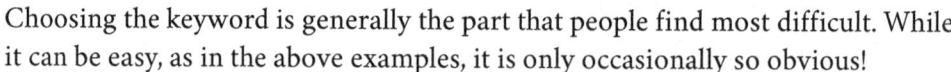

To remember that Canberra is the capital of Australia, you transform 'Canberra' into 'beer can', and put it in the hand of an Australian icon such as a kangaroo or a koala bear.

Choosing the keyword is generally the part that people find most difficult. While it can be easy, as in the above examples, it is only occasionally so obvious!

Good keywords are:

> ➤ acoustically similar to the word you're trying to remember (it's the sound of the word that's important, not the look of it)
>
> ➤ meaningful to you
>
> ➤ easy to visualize
>
> ➤ distinctive
>
> ➤ familiar (come readily to your mind)
>
> ➤ usually nouns or verbs rather than adjectives.

The keyword method includes *two* mnemonic links, that must both be remembered:

> ➤ the acoustic link (between the keyword and the word you're trying to learn — e.g., carta-cart; Canberra-beer can)
>
> ➤ the imagery link (between the keyword and the image — cart-image of the letter in the cart; beer can-image of the kangaroo swigging the beer can)

To use this method effectively, you need to practice both these links.

In general, the acoustic link is remembered very well. What tends to fail over time is the imagery link. Practice should therefore focus on this.

A good image is:

> simple

> clear and vivid

> interactive (a kangaroo drinking from a beer can, not a kangaroo and a beer can; a letter *in* a cart, not a letter and a cart)

> personally meaningful.

The 'imagery' link doesn't have to be a picture; it can be a verbal phrase.

The chief value of the keyword method is to help with **recognition** — to read 'carta' and recognize it as meaning 'letter'; to read 'Canberra' and recognize it as the capital of Australia. It is not as useful in the reverse task of recalling 'carta' when you want the Spanish word for 'letter'; of calling 'Canberra' to mind when you want the capital of Australia. However, you can improve its effectiveness for this by practicing the task in both directions.

The keyword method is far and away the most researched mnemonic strategy, and there is strong evidence for its effectiveness.

While mostly used to remember vocabulary (especially for learning another language, but also for learning new technical words in a subject), the keyword method has also been successfully used to remember simple associative facts, such as country or state capitals, authors of books, artists and their styles, attributes of minerals, or taxonomic information about animals.

So, for example, we could create keyword mnemonics to remember both the names of the early Americans and their associated areas:

Adenans — Ohio

Hohokam — Arizona

Adding nun

Santa ("**Ho ho**") with a **cam**era, taking a picture of the "**arid zone**"

Memorizing Verbatim

Hopewellians — Ohio

Mississippians / Temple Mound — Illinois

Is Miss Shirley Temple ill?

Anasazi — New Mexico/Arizona

A Nazi wearing a **Mexican** sombrero in the **arid zone**

First-letter mnemonics

Acronyms and acrostics are examples of first-letter mnemonics — memory strategies that use the initial letters of words as aids to remembering.

With acronyms, the initial letters form a meaningful word or pseudo-word (something that rolls off the tongue, even if it doesn't mean anything) — e.g., **ROY G. BIV** is an acronym for the colors of the rainbow; you could use **AHHMA** for the early Americans.

In acrostics, the initial letters are used as the initial letters of other words to make a meaningful phrase. "**R**ichard **O**f **Y**ork **G**ives **B**attle **I**n **V**ain" is an acrostic for the colors of the rainbow. "**A** **Ho**llow **Hope** **Miss**es **T**he **A**rgument" (with the 'T' reminding us of the alternate name for the Mississippian culture) is an acrostic for the early Americans.

Acronyms are very helpful indeed, and if the initial letters of some collection of items do happen to form a word, you should certainly take advantage of it. However, such opportunities are obviously rare.

Acrostics are much easier to create, although creating *good* ones is a little more challenging!

Principles for creating effective acrostics

Unfamiliar items need more cues. If the items are well-known, and the acrostic is only needed as a reminder or to provide order information, choice of words is only constrained by initial letter. However, if the items are not well-known, the words should also provide cues to the items.

Choose familiar words. Where possible, the words chosen should be familiar words (which are more easily recalled) rather than obscure ones.

Make it meaningful. As much as possible, the acrostic itself should make coherent sense (a meaningful sentence is remembered more easily).

Cue the order. If the acrostic is providing a particular type of order, where more than one type is possible, then it needs to also contain cues to what kind of order is involved.

Keep it simple. Don't force your mnemonic to carry too much information. If the acrostic is required to bear information about order, kind of order, *and* item content, it is usually better to create more than one mnemonic.

Usefulness of first-letter mnemonics

First-letter mnemonics are of limited value.

They are mostly useful when you need to remember the order of items (FACE for the notes on a stave; BEDMAS for the order of mathematical operations; "**M**y **V**ery **E**arnest **M**other **J**umps **S**even **U**mbrellas **N**ightly" for the order of planets).

They can also be used to help you remember the order of main points, for a speech or exam essay.

Their power lies in providing retrieval cues.

They are therefore also useful in fighting memory blocks, so are of particular value to those who suffer from test anxiety.

Watch out for the perils of repeated initials! This is the main source of errors when using such mnemonics.

The best time to use first-letter mnemonics is when:

- you have a relatively short list of items
- the items are themselves very familiar
- you need to remember the order of them
- the items all begin with different initials
- the items are related
- the items are concrete rather than abstract.

Simple list mnemonics

Most mnemonic strategies come under the heading of 'list' mnemonics — that is, like first-letter mnemonics, they are useful for remembering ordered information in particular, or any itemized list.

These tend to be primarily visual strategies, but the simplest of them has a verbal variant.

Story or sentence mnemonic

Consider the taxonomy of living things:

1. Kingdom
2. Phylum
3. Class
4. Order
5. Family
6. Genus
7. Species

Here's an attempt at a story:

In the KINGDOM, PHYLUM is a matter of CLASS, but ORDER is a matter for FAMILY, and GENIUS lies in SPECIES.

The trouble with this is not the re-coding of *genus* to *genius*; the trouble is, it doesn't make a lot of sense. It's a sentence, but not a story — there's no narrative. Humans think in stories. We find them easy to remember because they fit in with how we think. It follows then that the more effective story mnemonics will actually tell a story. To do that, we're going to have to transform our technical words into more common words.

King Phillip went to the **classroom** to **order** the **family genius** to **specifically** name the individual who had stolen the taxi.

The last part of this is of course unnecessary — you could finish it after individual if you wished. But an important thing to remember is that it's not about brevity. It's about memorability. And memorability is not as much affected by amount to remember, as it is by the details of what is being remembered. So meaningfulness is really important. Adding that little detail about stealing the taxi adds meaningfulness (and also underlines what this mnemonic is about: taxonomy).

Here's one for our early Americans: The **ad man ho-ho**ed, **hoping Miss Temple** would **answer**.

An effective story mnemonic

> tells a meaningful story

> uses familiar, preferably imageable, words

> includes elaborative details that help memorability

> is not too long (fewer than 9 items).

The story mnemonic is easily learned, but less effective than more powerful, but more difficult, list mnemonics.

Link mnemonic

The link method is the visual equivalent of the story method, using visual images (rather than words) to link items together. The link method requires less thematic coherence than the story method — you are essentially building a chain, in which the only requirement is that each item forms a visual image with the item next to it.

Advantages & disadvantages of the link method

> It doesn't require a pre-learned structure

> It is easily learned

> It doesn't allow you to go directly to any item, but requires you to work your way through the chain until you get to the item

> It requires visualization skill.

More complex list mnemonics

There are also more complex mnemonic strategies for remembering ordered information, and these are the mnemonics most often used and touted by memory champions. While effective, they do require a great deal of effort to master and use, and my own opinion is that few people actually want to put in the requisite time and effort for something that is rarely usefully employed.

What makes these strategies more effortful is the need to memorize (and practice to a high level) a specific structure — either a series of places or a series of numbered associations.

Method of loci

Also known as the journey method, place method, Roman room system.

You might use a familiar route, your house, or a particular room in it. The crucial thing is that you can easily call to mind various 'landmarks' (different fixed objects in a room, for example, or different buildings on a route). These landmarks are your anchors. You must train yourself to go around your landmarks in a particular order. With a route, of course, that is easy.

To remember a list, you simply visualize each item in turn at these landmarks.

Effective loci:

> - are very well-known
> - form a clear, circular route
> - are sufficiently dissimilar from each other not to be confused
> - are moderately, and reasonably evenly, spaced
> - are visualized in clear lighting
> - provide sufficient area for the arrangement of items.

In Roman times, this method was popularized as a strategy for helping people remember speeches.

Pegword mnemonic

The pegword mnemonic is based on the same idea as the place method, but uses numbers rather than places as cues. These numbers are transformed into visual images by means of the following simple rhyme:

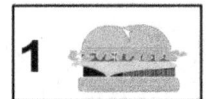
One is a bun

Two is a shoe

Three is a tree

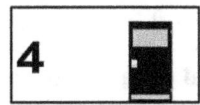
Four is a door

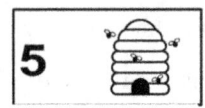
Five is a hive

Six is sticks

Seven is heaven

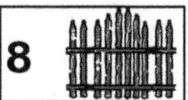
Eight is a gate

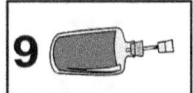
Nine is wine

Ten is a hen

Once you've got the rhyme firmly set in your head, you can associate items you want to remember with the visual cue:

The pegword method:

> requires you to learn the pegs very very well
> enables you to go directly to any item on a list
> is the best method for a numbered list.

Coding mnemonics

Most people find numbers much harder to remember than words. Accordingly, the best way to remember numbers or dates is to transform them into words. You can use any coding system you like, but here is the best-known one:

0 = s, z, soft c (*zero* starts with a *s* sound)

1 = t, d, th (there's *1* downstroke in *t*)

2 = n (*2* downstrokes in *n*)

3 = m (*3* downstrokes in *m*)

4 = r (*r* is the last letter of *four*)

5 = l (*l* is 50 in Roman numbers)

6 = sh, ch, j, soft g (*six* has a sort of *sh* sound)

7 = k, q, hard g, hard c (number *7* is embedded in *k*)

8 = f, v (both *8* and *f* have two loops)

9 = p, b (*9* is *p* the wrong way round)

Note that only consonants are used for coding. This means you can add vowels (and also, in this system, *w*, *h* and *y*) as necessary.

If you do want to learn this particular system, there is a mnemonic that may help you memorize the 0-9 codes: **Satan m**ay **r**e**l**i**sh c**o**ff**ee **p**ie. **Why** will help you remember which consonants can be used freely, like vowels.

The system also allows you to use doubled consonants where the sound doesn't change (which is mostly). For example, *dipper, dabble, squirrel*. Compare these to *accent*, where the first *c* is hard (7) and the second *c* is soft (0).

Similarly, a silent consonant doesn't count.

To remember the dates for our early Americans, then, we can encode

> 1000-200 B.C.E. as "**t**o**ss**es **n**ie**c**es **b**y" (the first two 's's would be better split up, but it's hard to get three separate 's's in a word, and you could argue that each is pronounced: tos-ses)
>
> 1-1450 as "**t**wo **tr**ai**ls**"
>
> 200-500 as "**n**ie**c**e**s l**ea**s**e**s**"
>
> 800-1500 as "**f**a**c**e**s t**wo **l**a**c**e**s**" (I'd prefer to have the second date in one word, but the rhyme helps make up for that)
>
> 900-1150 as "**b**a**s**e**s d**e**t**ai**ls**"

We can then incorporate these with our keywords to create:

> The adding nun tosses nieces by.
>
> Santa takes two trails.
>
> Hope for the nieces' leases.
>
> Miss Temple faces two laces.
>
> The nazi bases details.

Not brilliant, but that's how it works.

The coding mnemonic:

> ➢ requires the most time to master
>
> ➢ is an effective means of remembering numbers
>
> ➢ can be combined with the list-mnemonics to learn very long numbers
>
> ➢ can be combined with the pegword method to extend the number of pegs (using the codes to create concrete images that can become pegs).

Keeping it simple

Remember the codes with: **Satan m**ay **r**e**l**i**sh c**o**ff**ee **p**ie.

Vowels and consonants **why** don't count.

Avoid **x**, **ng**, silent consonants.

Remember doubled consonants that sound like one only count as one.

Remember it's the sound that matters.

Possible words for higher numbers:

1. **t**ie (man's necktie)
2. ho**n**ey (remember, h and y don't count!)
3. ha**m**
4. ea**r**
5. ow**l**
6. wi**tch**
7. **y**a**k**; e**gg**
8. **f**ae; i**v**y
9. **p**ie
10. **t**oe**s**
11. **t**oa**d**; **d**ea**th**
12. **t**i**n**
13. **t**i**m**e; **d**i**m**e
14. **d**ee**r**
15. **t**ow**l**
16. **d**i**sh**
17. **d**u**ck**
18. **d**o**v**e
19. **t**o**p**; **t**a**p**
20. **n**o**s**e

And so on, up to 99.

As you can readily appreciate, this enables you to greatly extend the pegword method. Moreover, although finding concrete words becomes increasingly hard as the number of digits increases, you can use a kind of hierarchical organization to extend your pegwords without having to come up with appropriate words. For example, by tying colors or other adjectives to particular groups, so that the nouns for numbers 100-199 are all described by one particular adjective, those for 200-299 by another.

However, I'm sure I don't need to point out that this strategy requires a LOT of work before you can use it. For most people, the effort won't be worth it.

Main Points

Mnemonic strategies are useful for helping you remember arbitrary details.

They are particularly helpful for remembering:

- the order of items
- technical or foreign words that have no meaningful connection to your existing knowledge
- simple, but arbitrary facts (such as how many days each month has)
- numbers.

Simple songs and rhymes are effective only when they're predictable.

Words can be as effective mnemonic aids as images — the trick is to make sure they're meaningful, and distinctive yet familiar.

Don't try to squeeze too much information into a single mnemonic.

Apply the right mnemonic to the information you want to remember. If the information doesn't fit one type of mnemonic, try another.

Don't feel you have to know them all!

- The keyword mnemonic is the most broadly useful and flexible (and not difficult to master).
- The coding mnemonic is the only choice if you need to remember dates and numbers.
- There are many list mnemonics, and it's better to become skilled at one or two, than know all of them not very well.

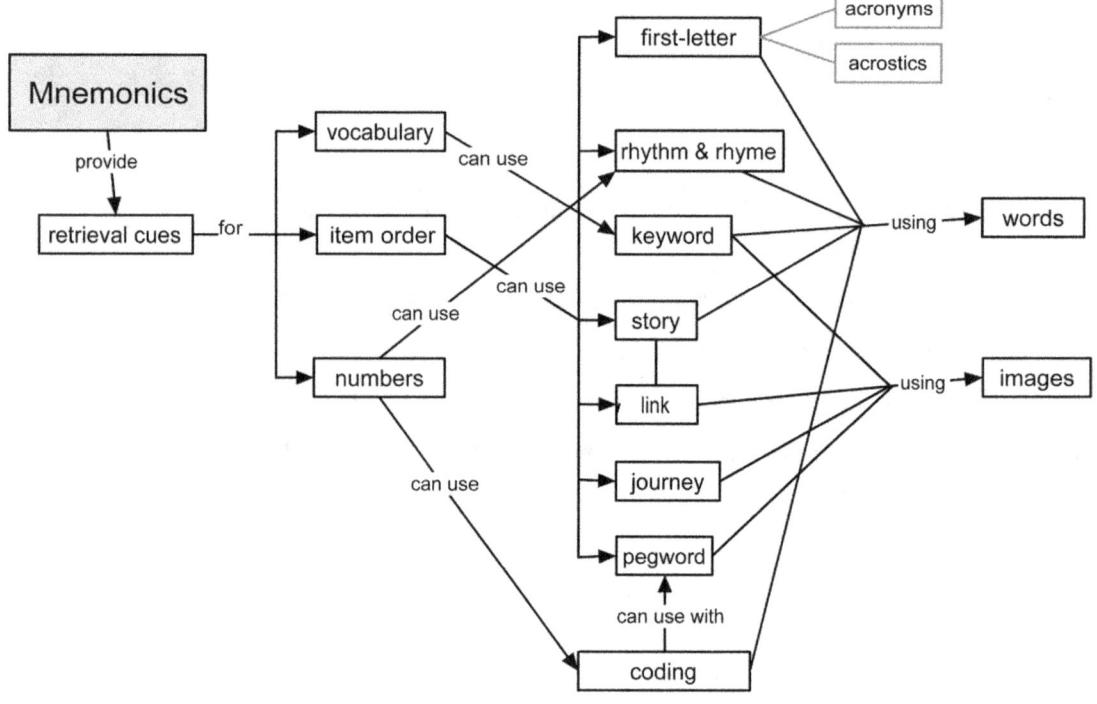

Review Questions

1. Mnemonics are useful for remembering:

    a. Your textbook

    b. The precise order of items

    c. Dates

    d. Vocabulary

    e. Concepts

2. Mnemonics are imagery-based. You shouldn't bother with them if you're not good at visualizing.

    a. True

    b. False

3. Rhythm and rhyme only help you remember when:

   a. They're catchy

   b. You can reliably predict what comes next

   c. They have an interesting rhythm

   d. They contain a lot of alliteration

4. The shopping cart in this image is:

   a. An imagery link

   b. An acoustic link

   c. A keyword

   d. The target word

5. What is the keyword in this image?

   a. Adding nun

   b. The numbers

   c. Oh Hi!

   d. The nun

Adding nun

6. Which of these is the acoustic link?

   a. Oh Hi!

   b. Adding nun

   c. Adding nun - Adenan

   d. Oh Hi! - Ohio

7. A good mnemonic image is:

   a. Funny

   b. Weird

   c. Interactive

   d. Simple

8. ROY G. BIV is:

   a. A first-letter mnemonic

   b. An acronym

   c. An acrostic

   d. A keyword

9. The journey method is useful for remembering:

   a. Places

   b. Lists

   c. Difficult concepts

   d. Speeches

   e. Main points in a speech

10. The pegword method is most useful for remembering:

    a. Lists

    b. Numbers

    c. Numbered lists

    d. Main points in a speech

Revising

Connections between memory codes are strengthened every time they're used. When they're not activated, they weaken. That's why repetition is so vital. But the sort of rote repetition we all learn to do as small children, is the least effective form of repetition there is.

To use your time and effort optimally, you need to follow the principles of effective practice.

The single most useful thing you can do to improve your learning is practice. But for this to be effective, you have to practice in the right way.

This is the general principle: **effective practice matches the task you want to master**.

Sometimes your goal is simply to be able to recognize the correct information. This is a much easier task, and **retrieval practice** is not necessary or even particularly helpful for this. But usually you will want to be able to retrieve the information from long-term memory. This means your revision should involve practicing retrieval.

Recall = retrieval

What is the wavelength of red light?

A question acts as a retrieval cue. To answer it, you must retrieve the information from long-term memory.

Recognition ≠ retrieval

Which of these is the wavelength of red light?

(a) 490 nm
(b) 630 nm
(c) 310 nm
(d) 850 nm

Picking the correct choice doesn't require you to retrieve the information, but simply to choose the most familiar one, or the one that 'feels' right.

Retrieval practice IS NOT repetition or rehearsal.

The idea is not to simply *repeat* the correct information, but to try to *retrieve* it.

Retrieval practice is a form of self-testing.

The effectiveness of retrieval practice for learning is the main reason why repeated testing is valuable (far more clearly valuable than repeated testing as a means of *assessment*).

To be effective, you must be **retrieving** from long-term memory, not working memory — therefore there needs to be sufficient time between retrievals for the information to clear from working memory.

When you are practicing or revising what you want to learn, think about how you will be needing to remember this in future. For example, if you're learning foreign language vocabulary, you may wish to be able to:

➢ remember the English meaning of the words when faced with them

- remember the foreign words when faced with their English counterparts
- spontaneously generate the foreign words when talking or writing.

What you practice depends on which of these tasks you want to be able to do.

Or perhaps you're learning about the French Revolution. Your aim may simply be to pass a multi-choice test at the end of the week, or it may be to get an A on an essay-type exam in two months time, or it may be that you actually want to remember most of these details for the long-term, and be able to talk intelligently about it as you incorporate them into your broader knowledge of history.

The first step in successful learning is always to think about what you want the learning *for*. Only then can you work out exactly how you should be learning it.

Retrieval practice:

- is the single most powerful learning strategy there is
- is a very simple, easily learned technique
- requires much less cognitive effort than other very effective strategies
- can be combined with other effective strategies, such as the keyword mnemonic or concept mapping (useful for difficult material)
- is most effective with feedback, but is still of some benefit even without feedback
- can even improve your memory for untested material — but only if it's related to tested information or appears in close proximity to it, and only if you spend some time searching your mind for related information when retrieving.

It has been suggested that retrieving incorrect information risks practicing the wrong information and thus making your retrieval of the correct information harder. This idea has lead to the '**errorless learning**' strategy, in which students are only given easy questions that they can readily answer correctly. However, this '**retrieval-induced forgetting**' only appears to occur in a narrow set of circumstances:

- when your learning — retrieval practice and testing — all occurs within a very short time period
- when you don't make any attempts to understand or elaborate the material to be learned

> when you practice by mixing up the order of your items (which is generally a good idea, but not when the first two factors are in play).

In other words, if you're practicing some rote-learned material (that is, information you haven't tried to understand, or memorize using mnemonic strategies), and you're practicing right before your test, then you shouldn't mix up the order of your items as you go through them. Otherwise, don't worry about this issue (although you do want to make a special effort not to repeatedly practice retrieving a specific piece of incorrect information).

There are three main ways to practice retrieval:

> flashcards
> Q & A
> re-summarizing.

Types of retrieval practice

Flashcards

Flashcards (or flashcard-type systems, such as making a vertical fold in a sheet of paper and writing items on one side and their answers on the other) are good for learning words and simple facts.

There are a number of flashcard software programs available, and the best of them use spaced practice principles. But if you are using actual flashcards, there are two important questions that affect your learning: how many cards should you study at a time, and when can you drop a card from the stack.

How many cards in a stack

A larger stack is much better than small stacks (this contradicts some advice, but this is what research tells us).

Twenty cards is probably a good starting point — increase the number if you're finding it too easy; reduce the number if it's too hard.

When to drop a card from the stack

Never drop a card after only one correct retrieval!

If you want to drop a card because you think you know it: give it one more turn. Students tend to drop cards too soon. **The value of studying is highest when items are closest to being learned.**

If you want to drop a card because you think it's too hard and you can't learn it: persevere. Have faith in the power of retrieval practice!

If it truly is difficult, apply an additional strategy (e.g., the keyword mnemonic).

If you drop too many cards, add more to the stack to maintain enough spacing between items (remember: you need the item cleared from working memory or there is no value in the retrieval).

Return dropped cards to the stack on the next review (though you can drop them sooner).

Q & A

To revise more complex and meaningful material, you need to produce a set of **learnable points**, which you can then turn into a 'Q & A' format.

Here's an example — a set of learnable points from the ozone text:

1. Ozone is important because it shields the surface from harmful ultraviolet radiation.
2. The stratosphere holds 90% of the ozone in our atmosphere (the ozone layer).
3. The troposphere holds 10%.
4. The troposphere is the lowest part of our atmosphere, where all of our weather takes place.
5. The ozone layer protects us; tropospheric ozone is a pollutant found in high concentrations in smog.
6. Wavelength is a measure of how energetic is the radiation.
7. The visible part of the electromagnetic spectrum ranges from 400 nanometers to 700 nm. Red light has a wavelength of about 630 nm; violet light about 410 nm.

8. Radiation with wavelengths shorter than those of violet light (at the short end of the visible spectrum) is called ultraviolet radiation. UV waves are dangerous because they're energetic enough to break the bonds of DNA molecules.

9. Of the three different types of ultraviolet (UV) radiation, the shortest (UV-c) is entirely screened out by the ozone layer, while the longest (UV-a) is not so damaging, so the main problem is UV-b.

10. The high reactivity of ozone results in damage to the living tissue of plants and animals, and is often felt as eye and lung irritation.

11. While our bodies can repair the damage done by UV waves most of the time, sometimes damaged DNA molecules are not repaired, and can replicate, leading to skin cancer.

12. The strong absorption of UV radiation in the ozone layer reduces the intensity of solar energy at lower altitudes. More energetic photons (ones with shorter wavelengths) are also less common.

13. Because ozone is most protective on the most dangerous wavelengths, a 10% decrease in ozone would increase the amount of DNA-damaging UV by about 22%.

14. Time and season affect how much UV radiation is absorbed by ozone because the angle of the sun affects how long the radiation takes to pass through the atmosphere (the path is shorter when the sun is directly overhead, so the radiation meets fewer ozone molecules).

15. Measurement:

    Solar flux = the amount of solar energy in watts falling perpendicularly on a surface one square centimeter; units are watts per cm^2 per nm.

    The action spectrum measures the relative effectiveness of radiation in generating a certain biological response (such as sunburn) over a range of wavelengths.

And here are the questions you might produce from such a set:

1. Why is ozone important?

2. What proportion of the atmosphere's ozone is in the stratosphere?

3. What proportion of the atmosphere's ozone is in the troposphere?

4. What part of the atmosphere is the troposphere?

5. What does wavelength tell us?
6. What is the range of the visible part of the electromagnetic spectrum?
7. What is the wavelength of red light?
8. What is the wavelength of violet light?
9. What does the ozone layer do?
10. What does tropospheric ozone do?
11. Where is this pollutant found?
12. Is ozone always protective?
13. Where is violet light in the spectrum?
14. What is ultraviolet radiation?
15. Why is it dangerous?
16. Which of the three different types of ultraviolet radiation is most dangerous and why?
17. How does ozone damage us?
18. Why does UV damage sometimes cause skin cancer?
19. What does the strong absorption of UV radiation in the ozone layer do?
20. Why is it safer for living things at lower altitudes?
21. Which wavelengths does ozone protect us from most?
22. How much would a 10% decrease in ozone increase the amount of DNA-damaging UV?
23. Why does time of day and season affect how much UV radiation is absorbed by ozone?
24. What is solar flux?
25. What unit is solar flux measured in?
26. What does the action spectrum measure?

How to create an effective Q & A for revision

Break each learning point down to the level which is right for you.

> One learning point may generate several questions.

> How much you break the learning point down depends on your background knowledge and understanding of the material.

Your questions should be neither too easy (if you already know the answer perfectly well, it's pointless to test it) nor too difficult (if you don't understand the material sufficiently well to answer the question, then you need to work on that more first).

Take care to frame questions so that they can stand alone and don't rely on your memory of the original sentence.

Put your questions in an order that makes sense to you, and keep that order for the initial reviews, only mixing it up when you are confident of your understanding.

Writing down your answers will be more effective than just saying them to yourself. Saying them aloud will be more effective than thinking them. Reading the question and thinking 'oh I know that', without actually articulating the answer, is NOT effective.

Re-summarizing

You can use any type of summary for review, but concept maps and mind maps are perhaps the most useful for retrieval practice, for several reasons:

> Many people find them more enjoyable than, say, writing down a list of points, or answering a list of questions.

> They can be a little different each time, giving you the opportunity to make new connections.

> They provide a spatial visualization — and spatial information tends to be more easily remembered.

I recommend using a concept map for initial review, when you're still consolidating your grasp of the material, and then perhaps using a mind map or simplified concept map (without labeling the links) on later reviews.

So, for example, your first review might produce something like this concept map (next page).

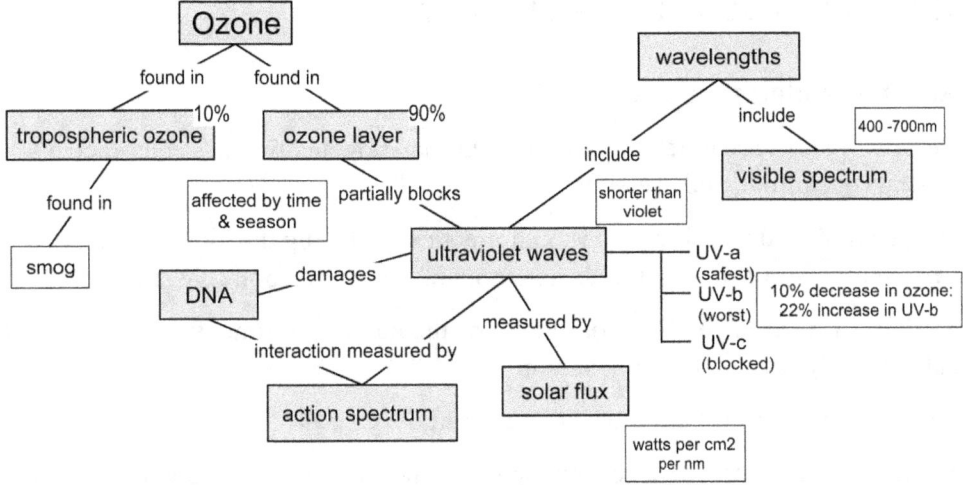

Your next review might produce something like this:

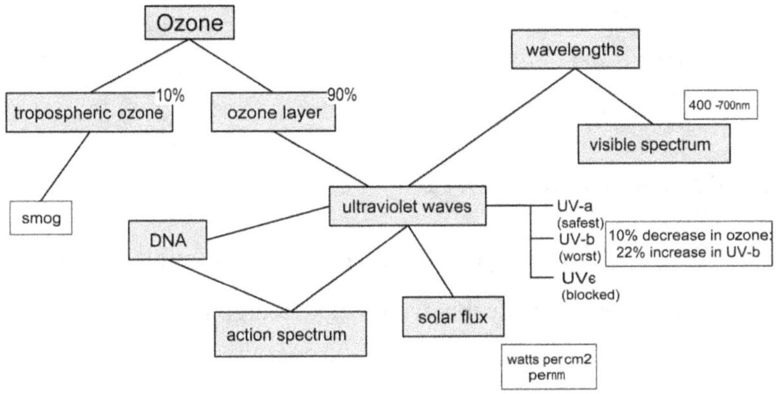

and your third review might be more like this:

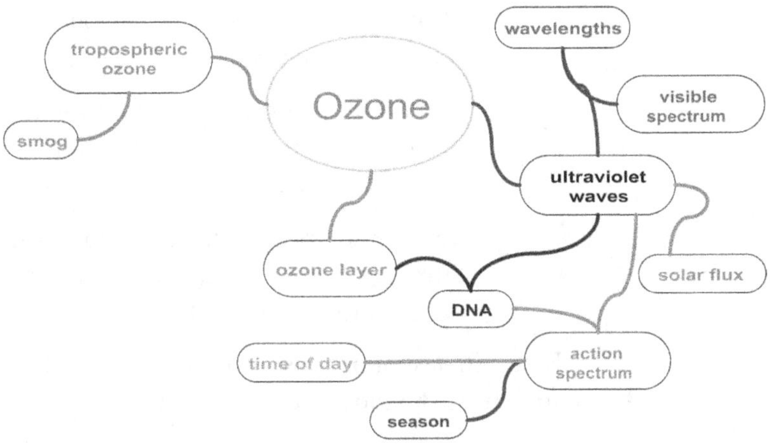

Please imagine this in lots of colors!

Revising

How often do you need to practice?

Recommended schedule

Review sessions are much more important than how many times you practice during your initial study session.

In your first study session, you should practice retrieving the information a certain number of times. Once is not enough! Three is a good rule of thumb.

Similarly, three is a good rule of thumb for the number of times you subsequently review the information.

Unsuccessful retrieval attempts do not count!

Because you want to avoid getting too bored, it's helpful to have some flexibility about the number of retrievals. Easy material can be limited to two successful retrievals, while harder material should have more.

Spacing your practice

Spacing within your study session

Within a study session, you can "mass" your practice, or space it out.

For example, if you're studying new vocabulary from a foreign language, you could:

> - test yourself on a word several times before moving on to the next, finishing your session when you come to the end of your words (**massed practice**), or
> - you could go through each word in the list, one by one, and then run through the list again, as many times as you think you need (spaced practice), or
> - you could repeat each word two or three times, in turn, then go through the list again, repeating each word two or three times again (**clustered practice**).

Research has found clustering is no more effective than massing. **Maximum spacing is the most effective strategy**.

What happens in the spaces

With individual items such as vocabulary, the 'spaces' between repetitions of any specific item are filled with other items in the list, as in the following example.

apple

spinach

pumpkin

lettuce

apple

spinach ...

Each repetition of the apple-der Apfel item is spaced by other items being practiced. Obviously here, the longer your list, the more space between repetitions. Too many items between, and you risk forgetting the item; too few, and you aren't getting the benefit of spacing (because it's still in working memory).

With category members such as different types of math problem, category members (rather than items) are interleaved:

1. If a prism base has 3 sides, how many faces will it have?

2. If a prism base has 5 sides, how many corners will it have?

3. If a prism base has 8 sides, how many edges will it have?

4. If a prism base has 6 sides, how many angles will it have?

5. If a prism base has 8 sides, how many faces will it have?

6. If a prism base has 4 sides, how many corners will it have?

7. If a prism base has 7 sides, how many edges will it have?

8. If a prism base has 3 sides, how many angles will it have?

In this case, 'category members' refer to variants that use the same formula. Thus, while questions 1 and 5 are different problems, they belong to the same category, that is, require the same formula.

Interleaving often makes learning seem more difficult — but while at the time of learning, you do worse than you would if you didn't interleave, a day or more later, you will remember it better than you would if you hadn't interleaved.

Interleaving and spacing are examples of **desirable difficulties** — difficulties that produce better learning. Desirable difficulties reflect a 'Goldilocks effect' — when the level of difficulty is 'just right', you'll learn better, *even though you don't seem to be learning well at the time.*

However, while interleaving can provide the right amount of difficulty, it also increases potential interference with your new memories.

Interleaving:

> - increases your cognitive load
> - increases interference
> - is of particular benefit in skill learning, math learning, and other types of category learning
> - is probably most helpful when you need to notice *differences* between interleaved items
> - is probably most useful when you've already achieved a certain level of competency with a skill or problem type or concept
> - can be helped by providing a brief rest during which the earlier learning can be stabilized (which reduces interference).

Interference is much less of a problem for pre-pubertal children, and much more of a problem for older adults.

For complex material, within-session spacing can be fruitfully achieved by providing breaks, rather than interleaving material.

Spacing between review sessions

Reviewing your material at least a day later makes a big difference to your chances of remembering it in future, compared to not reviewing it, or only reviewing it at the end of your study session.

The longer you want to remember the material for, the longer the length of time you should put between your initial study and your review.

Aim to review the material at the point where you still remember most of it, but will soon forget it.

As a guideline:

- ➤ Review after three days if you only want to remember for a week or two.
- ➤ Review after a week if you want to remember for a month.
- ➤ Review after two weeks if you want to remember for two months.
- ➤ Review after a month if you want to remember for a year.

For long-lasting learning, you need to review more than once. If this doesn't happen naturally in the course of your work or study, then you need to schedule some regular reviews.

Recommended:

- ➤ 3 reviews
- ➤ at increasingly longer intervals
  - ★ e.g., a first review one day after your initial study session, with a second review 7-10 days later, and a third review 4 weeks after that
  - ★ if you don't remember the material as well as you'd like, then shorten the next review interval and increase the number of reviews (i.e., don't count the poor review); if it's easy, stretch the interval to the next review
- ➤ for 'forever' learning, review the material after a year.

The evidence for the benefits of spaced over massed practice (both within a study session and between sessions) is overwhelming. After retrieval practice, the most important thing you can do to improve your learning is space your practice appropriately.

The ten principles of effective practice

1. Practice the task you need to do.
2. Practice retrieval from long-term memory.
3. When you practice retrieval, only correct retrievals count.
4. Aim to do at least two correct retrievals in your first study session.
5. Space your retrieval attempts out.
6. Review your learning on a separate occasion at least once.
7. Space your review out.
8. Review at expanding intervals for long-term learning.
9. Interleave your practice with similar material.
10. Allow time for consolidation.

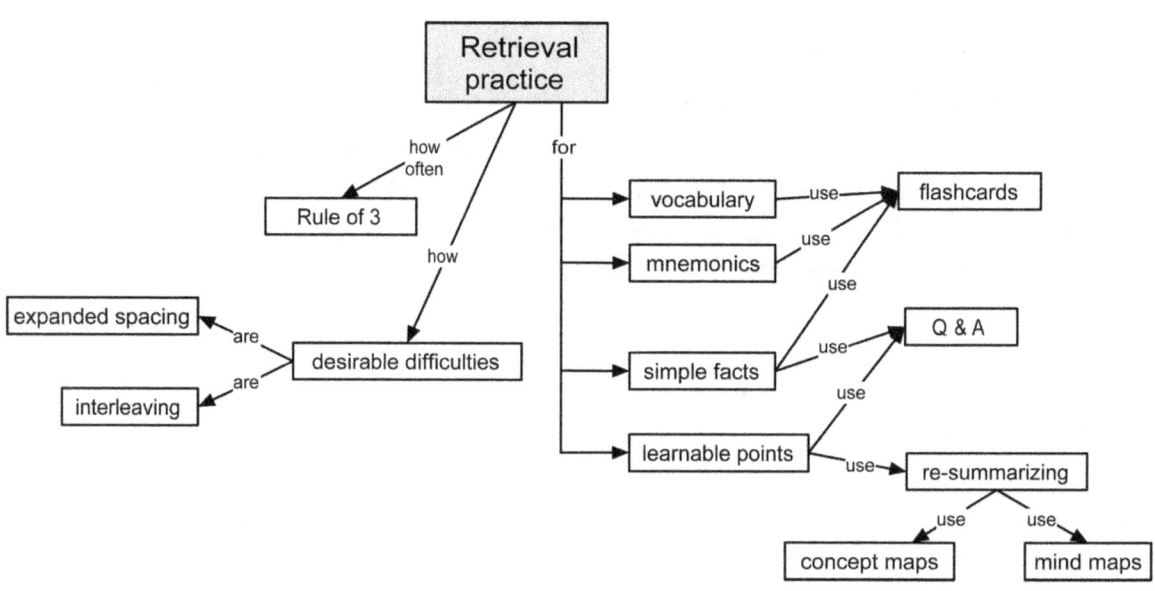

Review Questions

1. The most important principle you need to remember when revising is:

    a. Practice retrieval

    b. Space your retrieval practice

    c. Effective practice matches the task you want to master

    d. Practice everything three times

2. What's important to remember with retrieval practice is that:

    a. You should only ask questions you know you can answer correctly

    b. Only correct retrievals count

    c. You must have feedback on every retrieval

    d. You must leave a long enough gap between retrievals for the answer to leave working memory

3. Desirable difficulties are:

    a. Complications that add interest

    b. Complications that produce better learning

    c. Complications that make you work harder

    d. Complications that are at the level that is 'just right' for your knowledge and skill

4. Examples of desirable difficulties might include:

    a. Spacing

    b. Harder-to-read font

    c. Complex diagrams

    d. Interleaving

5. Clustered practice is:

   a. A good compromise between massed and spaced practice

   b. A poor compromise between massed and spaced practice

   c. Better than massed practice

   d. Better than spaced practice

6. The best way to revise is by repeating the material every day for 3 days:

   a. True

   b. False

7. Match each word list to the practice schedule it exemplifies:

A	B	C	
apple	apple	apple	Massed
spinach	apple	apple	Clustered
pumpkin	apple	spinach	Spaced
lettuce	apple	spinach	
apple	spinach	pumpkin	
spinach	spinach	pumpkin	
pumpkin	spinach	lettuce	
lettuce	spinach	lettuce	
apple	pumpkin	apple	
spinach	pumpkin	apple	

8. Effective revision is so important for long-term learning because:

   a. Repetition is the principal means of strengthening connections between memory codes

   b. It's hard to make memories stick if they aren't emotionally powerful

   c. New memories are vulnerable until they have been consolidated

   d. The connections between memory codes are only strengthened when retrieved

Skills

Skills, whether physical or mental, are learned quite differently from information. They are processed and stored in a different part of the brain, and this has implications for how they should be practiced.

The most important characteristic of skills is that they have the potential to be automatized. This is why practice is so utterly critical for skills, since automatization cannot occur without lots of practice.

Following the principles of effective skill practice will allow you to learn a skill significantly more quickly and efficiently.

Learning a skill is very different from learning information — something most of us are well aware of! But you may have thought less about what constitutes a 'skill'.

Study is not only about acquiring information; it's also about acquiring cognitive skills. Cognitive skills include reading, writing, study skills, performing mathematical calculations, speaking in a foreign language, problem solving, laboratory procedures, and more.

What makes a skill is its potential for automatization, which is only achieved through practice. Automatization means that you can perform the skill 'automatically', without conscious thought — meaning that it puts very little demand on working memory.

Practicing skills until they become automatic is therefore a great way of reducing the demands of a task (the cognitive load), of freeing up working memory space for those aspects that you do need to think about.

Critical factors in mastering skills

The 'gist' of actions is transferred very quickly to long-term memory, and remembered for a very long time even if not used. However, details are quickly lost, meaning that, if you want to keep a skill sharp, you need to keep using it. However, skills are readily refreshed when you return to them after an absence.

Learning a new skill will be easier if it shares elements with a skill you already have, but it will be more difficult if it contains elements that are antagonistic to an existing skill. For example, if you try to switch to a Dvorak keyboard having become skilled at typing on a Qwerty one; or if you try to write in an informal style after years of writing in an academic style.

Spacing and interleaving are particularly helpful for skill learning.

A necessary, though not sufficient, requirement for effective skill learning is regular **deliberate practice**.

Deliberate practice is focused and goal-oriented. In deliberate practice, you:

> - select a manageable chunk,
> - specify a goal,
> - monitor your performance, and
> - adjust it accordingly.

Goals should be responsive to your performance and needs, aimed either at specific steps/techniques, or at outcomes, as appropriate.

Variation, in your performance and in the environment, helps you develop good mental patterns that enable you to adapt to changing demands.

Deliberate practice is intensive and thus you're limited in how much you can do effectively at one time (rule of thumb: no more than an hour) and in one day (rule of thumb: no more than four hours).

Mental practice (imagining yourself performing the task) is helpful for cognitive skills, and particularly useful for complex tasks, and when you have some experience with the task (i.e., not so much for novices).

How you approach learning is more important than any specific strategy for working on your skill — you need to think analytically about your performance, monitor it, and respond constructively to errors.

With cognitive skills, deliberate practice is aimed at:

> learning the patterns important for the cognitive skill (such as text structures, or chess patterns), and

> automating sub-processes.

Because pattern recognition requires you to extract what matters from a mass of data, repeated experience is crucial, and a guide of some kind is extremely helpful.

Worked examples are the cognitive equivalent of physical demonstrations, and are most useful to novices. They can actually hurt your learning if you've already developed ways of doing the task that aren't consistent with the worked example. However, more experienced students can get more out of worked examples by explaining each step to themselves as they go.

Worked examples are also helpful for those with low WMC, because they help reduce cognitive load.

Effective worked examples:

> avoid splitting your attention (as happens when text and images are physically separated)

> avoid redundancy

> make subgoals (if any) explicit through labeling or by visually isolating sets of steps.

The ten principles of effective skill practice

1. Skill level is directly related to the amount of deliberate practice you do.

2. Break the skill down into self-contained sequences or sections that can be practiced separately.

3. Practice difficult sections more than easy ones.

4. Respond to errors by repeating the section/sequence more slowly and carefully, building up speed on successive repetitions.

5. Focus on only one aspect of the skill at a time.

6. Systematically vary the way in which you perform the skill, and the circumstances in which you do it.

7. Monitor and reflect on your performance.

8. Practice regularly but space it out — be aware of how long you can focus before your concentration fades, and allow time for consolidation.

9. Interleave your practice of sequences or problems with similar, but non-identical, sequences or problems.

10. Use observation of those who are better than you as a means of practice, and of learning new ways.

Review Questions

1. Examples of cognitive skills include:

    a. Reading a textbook

    b. Riding a bicycle

    c. Writing in French

    d. Long division

    e. Reading music

2. Learning Spanish and French at the same time is:

    a. Helpful, because they share a lot of similarities

    b. Difficult, because they contain elements that are antagonistic to each other

    c. More difficult than learning them at different times, because of the similar elements that are nevertheless different

3. Deliberate practice requires you to:

    a. Work hard, for as many hours as possible

    b. Break the task into smaller sequences

    c. Set clear goals and keep to them firmly

    d. Set clear goals and modify them as needed

    e. Work with all your concentration for as long as you can concentrate

4. Interleaving is best done using completely different tasks/topics:

    a. True

    b. False

5. The single most important characteristic of skills, both cognitive and physical, is:

   a. They're remembered very well

   b. They can be automatized

   c. They are maintained well without continuous practice

   d. You can practice them just by mentally imagining yourself doing them

6. Worked examples are most useful for:

   a. Experts

   b. Novices

   c. People with high working memory capacity

   d. People with low working memory capacity

Subjects

Different subjects require different ways of thinking, different ways of processing and presenting information. This is where the toolbox really comes into its own, because an effective student must modify their behavior according to the subject.

Although it's convenient to talk about "study skills", there is a danger in thinking of study skills as separate from the subject you're studying. Study skills, to a very large extent, are rooted in subject matter. Not only in the specific reading, organizing, and note-taking strategies you might use, but also in:

- your goals
- the way you manage your time
- your functional WMC.

Generalizing study skills, in the way I've done so far, is necessary to get the principles, but you must then apply these principles separately to your specific study areas. The most important message here is simply to be aware that you should treat different subjects differently! Don't read a chemistry textbook in the same way that you'd read a history text. Don't write a report for history in the same way you'd write one for English literature. Don't approach a math problem in the same way that you'd approach a biology problem. Don't do a presentation for psychology the same way you'd do one for media studies.

Learning how to apply general principles to specific subjects requires you to pay attention to what your teachers and textbooks tell you about how practitioners do things in that subject. Here are some examples to get you thinking.

Reading

Mathematics

Mathematics is an exact language and you can't read it for 'gist' — each word and symbol must be precisely understood.

In math, words often have specific meanings that are different from their general meanings. Such specific meanings need to be memorized. This can apply even in specific articles — mathematicians may memorize the variables being used before beginning to read, so that they don't have to go back and forth (this is also why it's important when writing a math paper to list the variables at the beginning).

Math textbooks often add a lot of extraneous information to their problems, in order to show relevance. Students may focus on this much more comprehensible and memorable material to the detriment of the details that are, in fact, what they need to learn.

Mathematicians, when reading proofs, focus on looking for errors.

Science

In science, reading is a transformative process — chemists, for example, write down formulas as they read text; mentally visualize; go back and forth between alternative representations (such as pictures, diagrams, graphs, text).

The main focus, when reading articles, is usually on understanding how an experiment was done and its result.

Again, many words used have both general meanings and meanings that are specific to the topic. It is vital, when reading science texts, to know what specific meanings the words have.

Like math, science textbooks also tend to add a lot of extraneous detail that, while more comprehensible and memorable, may mask the important details.

History

Historians, in contrast, pay particular attention to the source or author of a text as they read it, and try to work out what story the author is trying to tell. They realize they're reading an interpretation, not some absolute 'truth'.

Technical language is less of an issue in history.

Reading several sources on the same topic is critical.

Note-taking

The problem of extraneous detail can be handled by having specific structures for note-taking. So, for example, you might have a graphic organizer like this, for chemistry:

Substance	Atomic expression	Properties	Processes	Interactions

or this, for math:

Big Idea	Explanation	Definition	Example	Formula	Illustration

or this, for historical events:

What	When	Who	How	Why	Connections to other events

You can see here how the same type of structure can be used for very different information, simply by using very different headings.

Of course, some notetaking formats are more appropriate for some subjects than others (e.g., timelines for history; loop networks and tree networks for life sciences), but it's not so much the subject as the specific topic within it, that determines the best format. Across all subjects, then, text structure and your goals should be your guide.

Taking notes from books is generally discussed as if you have a single text. A useful strategy for dealing with the problem of reading and bringing together multiple documents (especially appropriate for history) is this:

1. read your first text
2. summarize it
3. read your second text
4. incorporate your summary into your previous summary, *keeping it much the same length*
5. read your third text
6. incorporate your summary into your previous summary, keeping it much the same length
7. and so on.

Writing

Academic reports and essays share some common requirements, such as

> assuming your reader has some knowledge of the subject but is not an expert

> defining abbreviations the first time you use them

> citing references according to whatever is the standard (this varies by subject area).

However, aside from such basics as these, the writing you are expected to do varies markedly by subject. For example (note that these are generalities, and specific subjects, journals, teachers, will have their own refinements):

Science report

Language: 3rd person, passive, past tense; use qualifiers (such as "possibly", "probably", "seems", "may", "might")

Quotes: rare

Preparation: experiment; literature review

Outlining: unnecessary, because the format is so prescribed.

Division: formal division into sections:

1. Title

2. Abstract: a one-paragraph summary, usually written in the present tense

3. Introduction: provides background and context for the experiment; includes a review of the relevant research (this is called a literature review); describes the broader significance of the research, the problem this experiment is attempting to answer, and your hypothesis.

4. Methodology / Materials and Methods / Experimental (exact label depends on the subject area or journal): provides sufficient detail for a reader (with the right expertise) to replicate your experiment; should not be a step-by-step account of what you did.

5. Results: presents the summarized data in a way that tells a comprehensible story to the reader; should include appropriate graphs and tables.

6. Discussion: this is where you interpret your results in light of the existing research and your hypothesis; describe any problems encountered; highlight any unexpected results and try to account for them.

7. Conclusions: a short 'wrap-up' summary, reiterating your goal, whether you achieved it, and the implications of the research.

8. Acknowledgements (if any)

9. References

Don't write the report in the order it will have when finished! Usually you start with the Methodology section, then Results, Discussion, Conclusions, Introduction, Abstract. References should be noted as you go, then tidied up at the end.

History report

Language: 3rd person, active, past tense; avoid qualifiers

Quotes: yes, but not too many

Preparation:

1. read several different documents on the topic
2. formulate a thesis
3. gather supporting evidence for your thesis (this may require seeking out additional documents)
4. note any contradictory evidence
5. note down any good quotes
6. select the most important points of evidence, both supporting and contradictory

Outlining: yes

- state thesis first
- note the points you'll make in your introduction
- note what background information you'll need to mention
- list each of your selected evidence points as a separate item

Division: not a formal division, but an expected order:

- Introduction
  - ★ arouse interest in first sentence
  - ★ state your thesis (this can be before or after your summary of evidence)
  - ★ summary of evidence
- Background
- Supporting evidence

- Contrary evidence (and why it's not as strong as supporting evidence)
- Conclusion: a brief summary, but it's main function is to discuss the broader implications of the topic
- References

Philosophy essay

Language: 1st person, active, present tense; minimize qualifiers

Quotes: avoid; if quoting, follow with your own paraphrase

Preparation: think a lot; read; discuss with others

Outlining:

- expect to make several outlines before you get it right
- expect to keep revising your argument as you write and think
- keep revising until the argument is completely clear — test it by explaining it to someone else
- look for the strongest contrary arguments you can find or think of — don't use weak arguments simply because they're easier to argue against

Division: not a formal division, but an expected order:

1. State your thesis
2. Define any technical or ambiguous terms that you use in the essay
3. Explain why this is an important or interesting thesis / problem — include relevant background information (such as what various philosophers have said), but keep it brief
4. Explain briefly the arguments you will make
5. Present your arguments — provide examples
6. Discuss contrary arguments — justify your claim that these are weak arguments
7. Conclusion: what you think your argument has established
8. References

Writing: expect to have to revise your draft several times

Understanding

'Understanding' is also a different process depending on the subject matter. For example:

- Understanding a foreign language text is a matter largely of translation — of knowing the meaning of the words and phrases, and the grammar that modifies and contextualizes them.
- Understanding a work of literature has to do with:
  - ★ knowing how story works
  - ★ knowing various other works for comparative purposes
  - ★ understanding human behavior.
- Understanding in history is a matter not of passively remembering what you read but of actively building an understanding from multiple texts, by:
  - ★ considering the source (the author or institution and their possible biases)
  - ★ comparing and contrasting the texts
  - ★ placing yourself in the context of the text's creation.
- In chemistry, understanding bonding and the size scale of the various constituents of matter (e.g., nucleus, atom, molecule, compound) is critical to your understanding of the subject.
- In math, understanding fractions is critical.
- In science, understanding includes connecting theoretical knowledge with familiar problems and experiences.

Remembering

The material you need to remember, and the strategies you employ, also depend on the specific subject. For example:

Foreign Languages

Vocabulary: flashcards; spaced retrieval practice; keyword mnemonic or other for difficult words

Grammar: explicit rules — spaced practice retrieving exemplars that vary across contexts

Fluency: deliberate practice

English literature

Emphasize narrative chain to anchor memory: practice retrieving by telling yourself or others

Main characters and important scenes: visualize; discuss with others; imagine the characters in different circumstances; reflect on different ways the scenes could have gone, what they revealed about the characters, etc.

History

Emphasize narrative chain to anchor memory: practice retrieving by telling yourself or others

Arguments, bringing together different sources: summarization; practice retrieving by arguing with yourself or others

Dates, names: mnemonics; spaced retrieval practice

Mathematics

Computational proficiency: deliberate practice

Problem-solving: deliberate practice of spatial visualization skills, of pattern recognition skills; gesture (gesturing as you think and discuss math problems has been found to reduce cognitive load)

Science

Correcting misconceptions: our 'folk knowledge' of how the world works is the biggest problem standing in the way of learning science; realizing how new facts contradict 'common sense' is the first and most important step in fighting this.

Making connections: asking questions; making comparisons; creating concept maps

Spatial visualization: deliberate practice

Perceptual learning / pattern recognition: deliberate practice

Memorizing technical jargon and arbitrary facts: mnemonics; spaced retrieval practice

Main Point

That different subjects can't all be treated the same!

This is only a brief overview, intended to direct your attention to some of the ways in which subjects are different, requiring different approaches and strategies. Use this as a guide, not a rule-book.

Review Questions

1. When you read a textbook, you should always pay attention to the precise meaning of technical words:

    a. True

    b. False

    c. It depends on the subject

    d. It depends on the subject and your goal

2. Writing a good report or essay depends on following a very specific format:

    a. True

    b. False

    c. It depends on the subject

    d. It depends on the subject and your goal

3. Concept mapping is an excellent strategy for understanding a topic:

    a. Always

    b. For some topics

    c. In science only

    d. For some topics more than others

Putting it All Together

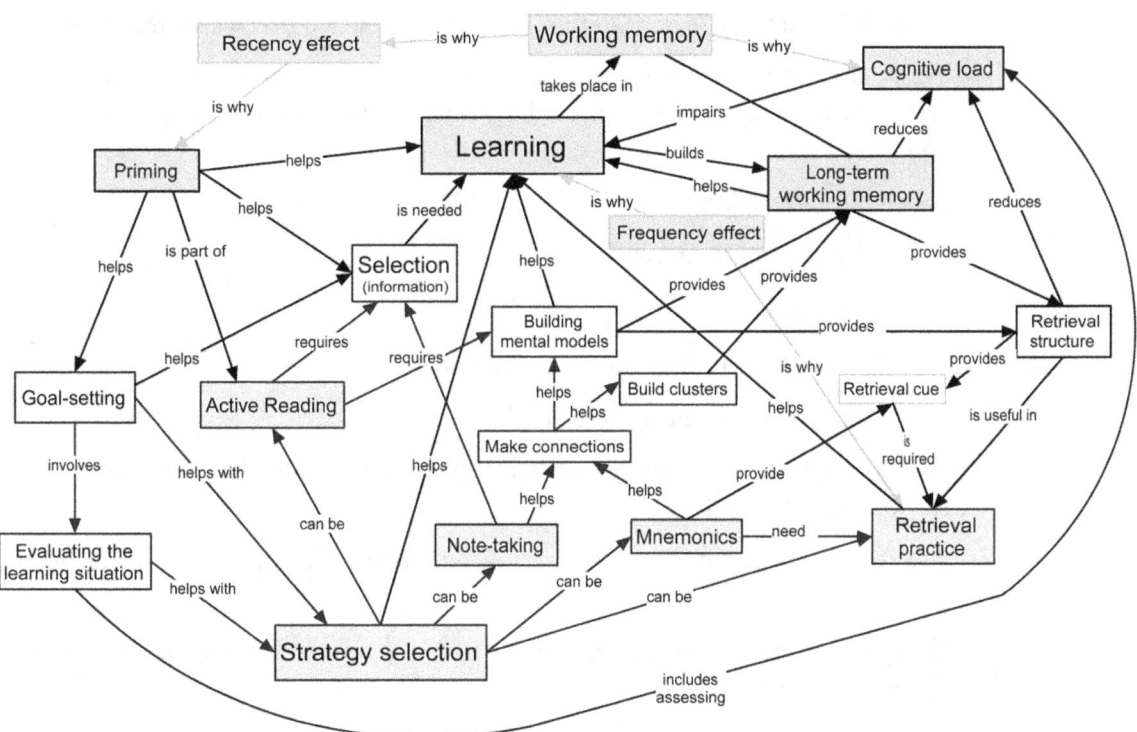

The main steps

Prime your brain for learning, and for the subject matter.

Set your goals, for the material and for this study session.

Consider the material and the task: is it a lecture, a text, a hypermedia situation?

Get a preliminary 'big picture' view of the material you're going to cover:

- skim the text
- skim handouts and recommended reading (lectures)
- study the navigation (hypermedia).

Process the material (e.g., read the text; listen to the lecture; watch the videos).

Think about the structure of the material and the best way to format / organize your notes (if a lecture, your 'recording' notes will of course need to be made as you listen).

Start your note-taking.

Make clear distinctions between

- skills
- information that must be memorized
- material that needs to be understood.

Put information to be memorized in an appropriate format (e.g., flashcards; learnable points for Q & A).

Use retrieval practice for memorization, augmented by mnemonics for difficult items.

Use questioning and concept mapping strategies to help you understand more complex material.

Practice retrieval of questions and maps to help you remember the material.

Use deliberate practice to build your skills, trying to reach automatization where possible (in simple skills, or in components of more complex skills).

Remember that note-taking, reading, and revising are processes that usually need to be repeated, modifying your approach or focus each time.

Your toolbox

In this book, I have tried to give you a toolbox of effective strategies, and the understanding to use that toolbox effectively. Remember that you are an individual, with your own ways of thinking, your own preferences, your own goals. Don't let anyone tell you there is only way to learn! This book is about showing you some of the choices available to you. Make your own toolbox, with the strategies that work for you, in the circumstances you find yourself in. And remember that circumstances change, and so do you. Let your toolbox change then too.

I wish you well on your journey.

Glossary

acronym: uses the initial letters of a list of items to form a meaningful word (e.g., FACE for the notes in the spaces of the treble staff)

acrostic: uses the initial letters of a list of items as the initial letters of other words to make a meaningful phrase (e.g., Every Good Boy Deserves Fruit for the notes on the lines of the treble staff).

active reading: strategies that promote the effective selection of important information by encouraging active involvement in the reading process.

advance organizer: information that appears before the text it refers to, with the purpose of putting the information in a broader context. Advance organizers are written at a higher level of abstraction than the text, and may be written in a graphical format.

analogy: a type of comparison that helps you understand something; uses similarity of structure, of relations between objects.

anchor points: facts that you already know very well, that connect to information you're trying to learn. By tying new information to information you already know, anchor points help make the new information meaningful and more memorable.

chunk: a tight cluster of information able to be treated as a single unit when worked with.

clustered practice: a hybrid of massed and spaced practice; no more useful than massed practice.

code principle: every memory is a selected and edited code, not a recording of real-world events.

coding mnemonic: a mnemonic strategy for transforming numbers into letters, to make them easier to remember.

cognitive load: the burden on your working memory system made by information-processing tasks.

coherence gap: gaps in the text that the reader is expected to fill in from their background knowledge.

consolidation: the process of further editing and stabilizing new memories for long-term storage.

context effect: the degree to which the context in which you are trying to retrieve information matches the context in which you originally encoded it affects how easy it is to retrieve.

context: the information contained in the situation in which you are encoding or retrieving the target information. It includes the physical environment and your own physical, mental and emotional state, as well as information presented at the same time as the target.

correspondence list: a list of specific correspondences between the elements of two analogical examples that are being compared — a step in the mutual alignment strategy.

cue recognition: recognizing the cues in the text or lecture that indicate what information teachers or writers regard as important.

cued recall: remembering in response to prompts, such as specific questions.

deliberate practice: is a means of achieving expertise in skills that requires intense focus and constant monitoring.

desirable difficulty: a degree of difficulty, such as reading in a hard-to-read font, that encourages learners to put more time and effort into processing the information, resulting in better learning.

distinctiveness principle: memory codes are easier to find when they can be easily distinguished from other related codes.

domino principle: activating one memory code causes other, linked, codes to be activated also.

elaboration: any strategy that involves expanding on the information presented — the aim usually being to connect new information with familiar information. It does not necessarily involve deepening your understanding of the material; mnemonic strategies can involve elaboration.

elaborative interrogation: a strategy that attempts to bring to mind relevant prior knowledge you may have, by asking yourself why the new information is true.

encoding: the process of transforming information into a memory code, and placing it in your memory.

errorless learning: the idea that students learn better if they're given easy questions and problems that they're unlikely to get wrong.

expository text: text whose main purpose is to provide information. It contrasts with narrative text, whose main purpose is to tell a story.

first-letter mnemonic: is a list-learning strategy that uses the initial letters of the items to aid recall. There are two types: acronyms and acrostics.

free recall: remembering without prompts — for example, when you're asked to write an essay on a topic.

frequency effect: the more often a code has been retrieved, the easier it becomes to find.

goal-setting: your articulation of the target at the beginning of your search. The more specific it is, the more likely your search will be successful, and the more likely you are to recognize that you have indeed reached your target.

graphic organizer: a type of graphic summary appropriate for material that can be expressed hierarchically, that allows the comparison of between-cluster relations. Common examples are tree diagrams and matrix diagrams.

graphic summary: a summary that re-organizes the information in a more visual format. Examples are graphic organizers, outlines, multimedia summaries, maps.

headings: single words or phrases that label sections in a text and help organize it in a hierarchical structure.

highlighting: any way of emphasizing key words or phrases, such as underlining, framing, using bold type, or using a colored marker.

hypercorrection effect: when students are more confident of a wrong answer, they are more likely to remember the right answer if corrected.

hypermedia environment: a non-linear multimedia environment.

imagery: the use of visual images to encode non-visual information.

interleaving: interspersing practice of one type with practice of other types.

joint interpretation strategy: a statement describing two analogical examples as a unit — a step in the mutual alignment strategy.

journey method: see *method of loci*

keyword method: transforms a word into an image via a keyword — a word derived from the word to be learned, that is imageable. The keyword method is useful for linking pairs of items — a word with its meaning, a capital with its country, a country with its product.

learnable point: important information expressed concisely in a statement that can be easily turned into a question-&-answer format.

link method: has no well-learned anchors, but simply links items in a chain of paired items.

list-learning strategies: include three strategies using imagery to link items in a list: the place method, the pegword method, and the link method, and two strategies using words to link items: the story method, and first-letter mnemonics.

long-term working memory: information that is recently out of working memory can be kept readily available if it is part of a strong network of linked concepts that is linked to information in working memory.

maps: graphic summaries that display main ideas in a structured but non-hierarchical format.

massed practice: the opposite of spaced practice — concentrating your learning in a solid block.

matching effect: a memory code is easier to find the more closely the code and retrieval cue match.

mind-mapping: a mapping strategy made famous in a number of books by Tony Buzan.

mnemonic: aids to memory such as acronyms, acrostics, and techniques that link information by creating visual images or making up a story. They are most suitable for information that is not inherently meaningful.

multimedia summary: a graphic summary that combine pictures and text in an integrated manner. Especially appropriate for demonstrating scientific explanations.

mutual alignment: A way to find analogies, by comparing two partly understood situations, searching for the common structure. This compares to the more familiar strategy of comparing an unfamiliar situation with a familiar one (which requires you to be familiar with a situation that is analogous).

network principle: memory consists of links between associated codes.

organizational signals: are devices that highlight the topic structure of a text, such as headings, overviews and topical summaries.

outcome goals: your objective in carrying out a learning task, in terms of the desired outcome. This contrasts with process goals.

outline: a type of graphic summary that systematically lists concepts with their subordinate concepts and their attributes.

overview: a topical summary that appears before the text it is summarizing.

pegword method: is similar to the place method, but uses numbers as pegs or anchors. Images for the numbers are rote-learned by means of a rhyme. The pegword method can be extended using a coding system.

method of loci: also known as the journey method, or Roman room system. Images of items to be remembered are visualized at familiar landmarks, in a set order.

priming effect: a memory code is readier to activate, and so easier to access, when memory codes linked to it have been recently activated.

priming: refers to the process of preparing a system for functioning — in this context, preparing the brain for learning by activating relevant prior knowledge and directing attention appropriately.

process goals: specific intermediate objectives that need to be achieved on the way to producing the desired outcome of a learning task.

recall: the retrieval of information from long-term memory.

recency effect: a memory code is more readily activated when it has recently been activated.

recognition: the awareness that you have seen or learned this information before. Multi-choice tests assess recognition rather than recall.

reconsolidation: stable memory codes become labile again (capable of being changed) after reactivation, suggesting that consolidation, rather than being a one-time event, occurs repeatedly every time the representation is activated (that is, retrieved from long-term memory).

retrieval context: the situation in which you attempt to remember the information. In the study situation, examples include an exam, multi-choice test, classroom discussion, writing an essay, or a brainstorming session.

retrieval cue: something that prompts you to recall a specific memory.

retrieval practice: the strategy of repeatedly trying to retrieve the information to be learned.

retrieval-induced facilitation: when retrieval practice improves memory for related, untested information.

retrieval-induced forgetting: when retrieval of information blocks the retrieval of other information.

retrieving: finding a memory code; 'remembering'.

Roman room system: see *method of loci*

self-explanation: a strategy that involves you explaining the meaning of information to yourself while you read.

self-monitoring: strategies to inform you how well you have learned the information in a memory situation so that you can plan your encoding strategies appropriately.

sentence mnemonic: see *story mnemonic*

skimming: skipping speedily through text actively searching for critical information.

spaced practice: reviewing learning or practicing a skill at spaced intervals, rather than in one concentrated block.

stabilization: the first stage of memory processing, lasting about six hours, during which new information is particularly vulnerable to being lost.

story mnemonic: links words to be remembered in a simple story told in one or two sentences.

text structure: the underlying organization of a text.

topic structure strategy: a strategy for processing text that involves reorganizing the information to reflect the presumed hierarchical organization of topics and sub-topics that underlies the passage. The strategy is contrasted with the listing strategy.

topical summary: is a simple factual summary of the main points of a text that doesn't add any new information or offer a new perspective (for example, by re-organizing the information).

working memory: includes the part of memory of which you are conscious; the "active state" of memory. Information is held in working memory during both encoding and retrieval. Working memory governs your ability to understand, to learn new words, to plan and organize yourself, and much more.

working memory capacity: the amount of information you can hold and work with at one time. Now thought to be 3-5 chunks.

Answers to Review Questions

How Memory Works
1. B, C, D
2. A, B, C, D
3. C
4. D
5. A, B, C

Approaching your Learning
1. B
2. B
3. C
4. B, C, D
5. A, B, D
6. B
7. D
8. A

Getting an Understanding of the Text
1. D
2. A, B, D, E
3. A, B, E
4. C, D
5. A, B, D, E, F
6. A, B, C, D
7. A, B, C, D
8. A, B, C, D

9. A—representation

   B—organization

   C—interpretation

   D—decoration

   E—transformation

Reading

1. A, B, C, D
2. A, B, D, E
3. C
4. D

Taking Notes

1. B
2. C, D, E
3. C
4. A, D
5. B, C, E
6. E
7. A—Cause-&-effect

   B—Comparison

   C—Description

   D—Description

   E—Refutational
8. D

Learning through Understanding

1. B, C, D, E
2. E

3. Atom : Solar system

   Nucleus : Sun

   Electrons : Planets

4. D

5. A

Reading on the Web

1. B, C, E
2. A, C, D
3. A, D
4. B, E
5. A, B, C, D, E

Getting the Most out of Lectures

1. B
2. B
3. C
4. A, B

Memorizing Verbatim

1. B, C, D
2. A
3. A, B
4. C
5. A, C
6. C, D
7. C, D
8. A, B
9. B, E
10. C

Revising

1. C
2. B, D
3. B, D
4. A, B, C, D
5. B
6. B
7. A—Spaced
   B—Massed
   C—Clustered
8. A, B, C, D

Skills

1. A, C, D, E
2. C
3. B, D, E
4. B
5. B
6. B, D

Subjects

1. D
2. C
3. D

www.ingramcontent.com/pod-product-compliance
Lightning Source LLC
Chambersburg PA
CBHW081719100526
44591CB00016B/2432